Noise Control of the Beginning and Development Dynamics of Accidents

Telman Aliev

Noise Control of the Beginning and Development Dynamics of Accidents

Springer

Telman Aliev
Institute of Control Systems
Azerbaijan National Academy of Sciences
Baku, Azerbaijan

ISBN 978-3-030-12511-0   ISBN 978-3-030-12512-7 (eBook)
https://doi.org/10.1007/978-3-030-12512-7

Library of Congress Control Number: 2019930990

© Springer Nature Switzerland AG 2019
This work is subject to copyright. All rights are reserved by the Publisher, whether the whole or part of the material is concerned, specifically the rights of translation, reprinting, reuse of illustrations, recitation, broadcasting, reproduction on microfilms or in any other physical way, and transmission or information storage and retrieval, electronic adaptation, computer software, or by similar or dissimilar methodology now known or hereafter developed.
The use of general descriptive names, registered names, trademarks, service marks, etc. in this publication does not imply, even in the absence of a specific statement, that such names are exempt from the relevant protective laws and regulations and therefore free for general use.
The publisher, the authors and the editors are safe to assume that the advice and information in this book are believed to be true and accurate at the date of publication. Neither the publisher nor the authors or the editors give a warranty, express or implied, with respect to the material contained herein or for any errors or omissions that may have been made. The publisher remains neutral with regard to jurisdictional claims in published maps and institutional affiliations.

This Springer imprint is published by the registered company Springer Nature Switzerland AG
The registered company address is: Gewerbestrasse 11, 6330 Cham, Switzerland

Preface

It is known that the causes of the transition of objects to the latent period of an emergency state are associated with the emergence of various defects. They are reflected in the form of noises on the signals received at the outputs of object's corresponding sensors. At the same time, these noises correlate with the useful signal and contain valuable diagnostic information that is lost when traditional processing technologies are used. Therefore, they do not allow to control the beginning and dynamics of the development of accidents reliably and adequately.

For different objects, the process of defect initiation and development to a pronounced emergency state has distinctive features that can be related to object's physical, biological, mechanical, chemical, and other properties, or the functions that they perform, their operation modes, etc. For these and many other reasons, defect initiation to a pronounced emergency state proceeds in different ways in different objects. In some objects, this happens quickly. In others, this process takes much more time. However, despite all these differences, they have in common that in that period, changes in the state of the object manifest themselves in the signals in the form of noise that contains valuable diagnostic information. At the same time, development dynamics of accidents in their latent period affects the correlation time and the value of the estimate of the cross-correlation function between the useful signal and the noise. For this reason, solving the control problem in the latent period of object's transition to an emergency state differs from the diagnostics of the technical condition of objects during their operation in normal mode. Therefore, it is often impossible to obtain reliable results and generate solutions that would be adequate for the situation by using relevant control and management systems. This is often the cause of numerous accidents at various oil and gas production facilities, petrochemical, power engineering facilities, aviation objects, etc., with human casualties and catastrophic financial and environmental consequences. And in these systems, in many cases, filtering methods are used to eliminate the effect of noise on the results of the problems being solved. At the same time, noise often emerges as a result of the transition of control objects to an emergency state and becomes the carrier of information that is erased by filtering. As a result, the important, in some cases unique, valuable information about the beginning of the latent period of the emergency state is lost.

Thus, traditional technologies for analyzing noisy signals provide no way of extracting and accounting for the information contained in the noise of analyzed signals. This is a serious reason behind the difficulty of control of objects' operation in the latent period of an emergency state. And this, in turn, makes it difficult to ensure their accident-free operation. Therefore, it is necessary and appropriate to create technologies and intelligent noise control systems that have the ability to control the beginning and development dynamics of accidents in their latent period.

Detection of the process of initiation and development of a crack, wear, deformation caused by fatigue or overload, etc., in the early stages is of enormous importance for many modern critical facilities (turbines, offshore platforms, aircrafts, tankers, etc.). For instance, when a microcrack forms in metal trusses and supports of a fixed offshore platform, one first hears faint squeaking, which then grows louder, then it reminds of a baby's crying, turns into metallic creaking, etc. In this case, the noise of the acoustic signal also becomes the carrier of information about the initiation of the defect.

It is obvious that after a defect has been detected, for many objects, its development dynamics must be controlled to be able to estimate equipment's remaining life of before and the time for routine or overhaul repairs.

It is evident from the above that the state of each object changes after a certain period during its operation. First, in the period T_0, it is in its normal state. Then, during the period T_1, the object goes into the latent form of an emergency state. After this, it goes into the period T_2 of the emergency state. Then the period T_3 of the pronounced emergency state begins, when an accident occurs and the object stops functioning.

Despite the differences in the duration of the periods T_0, T_1, T_2 for different objects, the control problem comes down to providing a reliable indication of the beginning of time T_1 of the latent period of changes in the state of the control object.

However, in the known control and diagnostic systems, mainly due to the above reasons, object's emergency state is registered at the beginning of the time period T_2. Because of this, the information about object's transition to an emergency condition in some cases is obtained too late. Therefore, to control the beginning and development dynamics of accidents at the beginning of the time period T_1, it is advisable to apply the technology and system of noise control.

Noise control of the beginning of accidents implies early detection of defect initiation, in the state at which defect's effects on the reliability or operability of the equipment are not yet manifested. In the first place, it is necessary to control those defects and faults that can cause catastrophic accidents.

On the basis of the noise technologies proposed in this book, various intelligent systems have been developed, designed, and implemented for the noise control of the beginning and development dynamics of accidents at oil and gas production facilities, construction sites, in power engineering, transport, seismology, aviation, etc. They also make it possible to improve the accuracy of the results of traditional correlation and spectral technologies for analyzing noisy random signals, as well as the adequacy of solutions to numerous recognition, identification, modeling, and control problems.

Baku, Azerbaijan   Telman Aliev

Contents

1 Specifics and Difficulties of Control of the Beginning of Accident Initiation and Development Dynamics 1
   1.1 Causes and Specifics of Accident Initiation 1
   1.2 Types and Development Stages of Defects Preceding Accidents in Technical Facilities 3
   1.3 Sensors and Specifics of Information Support for the Control of the Beginning of Accident Initiation and Development 6
   1.4 Models of Signals Received at the Sensor Outputs in the Latent Period of Accidents 9
   1.5 Difficulties in Controlling the Beginning of the Latent Period of Accidents with the Use of Traditional Technologies 12
   1.6 Factors Affecting the Adequacy of the Control of Accident Initiation by Correlation Analysis Methods 13
   1.7 Factors Affecting the Adequacy of the Control of Accident Initiation by Spectral Analysis Methods 16
   1.8 Effects of the Signal Filtration on the Result of the Control of the Beginning of Defect Initiation 19
   1.9 Effects of Traditional Methods of Sampling Interval Selection on the Adequacy of Control of the Beginning of Accident Initiation 20
   References 22

2 Correlation Technology for Noise Control of the Beginning and Development Dynamics of the Latent Period of Accidents 27
   2.1 Correlation Noise Technology for Calculating the Variance of Noise of Noisy Signals in the Latent Period of Accidents 27
   2.2 Correlation Technology of Noise Control of Accident Development Dynamics 33
   2.3 Correlation Technology of Noise Signaling for the Beginning and Development Dynamics of the Latent Period of Accidents ... 36

	2.4	Technology for Analyzing the Estimate of the Density of Distribution of Noise by Its Equivalent Samples	40
	References	43

3 Algorithms for Forming Correlation Matrices Equivalent to Matrices of Useful Signals in the Latent Period of Object's Emergency State 45
   3.1 Difficulties in Forming the Correlation Matrices in the Latent Period of Object's Emergency State 45
   3.2 Technologies for Forming the Equivalent Correlation Matrices for an Object in the Normal Operating Mode 49
   3.3 Technology for Forming the Equivalent Correlation Matrix in the Latent Period of Object's Emergency State 52
   References 56

4 Spectral Technology for Noise Control of the Beginning and Development Dynamics of the Latent Period of Accidents 59
   4.1 Algorithms and Technologies for Calculating the Errors in the Estimates of Spectral Characteristics in the Latent Period of Object's Emergency State 59
   4.2 Algorithms for Calculating the Estimates of Spectral Characteristics of the Noise in the Latent Period of Object's Emergency State 62
   4.3 Spectral Technology of Noise Signaling for the Beginning of the Latent Period of Accidents 66
   4.4 Position-Binary Technology for the Control of the Beginning of the Latent Period of Accidents of Objects of Periodical Operation 68
   4.5 Position-Selective Technology for Calculating the Sampling Interval of the Noise in the Latent Period of Object's Emergency State 74
   References 77

5 Application of Technology and System of Noise Control on Fixed Offshore Platforms and Drilling Rigs 79
   5.1 Systems for Noise Control of the Beginning of the Latent Period of Accidents on Fixed Offshore Platforms 79
   5.2 Technologies and System for the Noise Control of the Beginning and Dynamics of the Development of Accidents on Drilling Rigs 88
   References 100

6	**The Use of Noise Control Technology and System at Oil and Gas Production Facilities**	101
	6.1 Noise Control System of Sucker Rod Pumping Unit	101
	6.2 System for Noise Control of the Beginning of the Latent Period of Accidents at Compressor Stations	115
	References	119
7	**Possibilities of the Use of the Technology and System of Noise Control of the Beginning of the Latent Period of Accidents on Technical Facilities**	121
	7.1 Specifics and Relevance of the Use of the Technology and System of Noise Control at Power Engineering Facilities	121
	7.2 Intelligent Robust System for Noise Control of the Latent Period of Transition of Power Engineering Facilities into an Emergency State	126
	7.3 Subsystems for Noise Control of the Beginning of the Latent Period of Nuclear Power Plant Equipment's Transition into an Emergency State	129
	7.4 Possibility of Preflight Noise Control of the Technical Condition of Aviation Equipment	134
	7.5 Possibilities of Using Noise Control Technology in Railway Safety Systems in Seismically Active Regions	139
	References	143
8	**Using Noise Control Technologies and Systems in Construction and Seismology**	145
	8.1 Technology and System of Noise Control of Anomalous Seismic Processes	145
	8.2 Digital Citywide System of Noise Control of the Technical Condition of Socially Significant Objects	155
	8.3 Intelligent Seismic-Acoustic System for Identifying the Area of the Focus of an Expected Earthquake	158
	References	178
9	**Possibilities of Using Noise Control Technology in Medicine**	181
	9.1 Possibilities of Using Laptops and Smartphones for the Control of the State of the Heart	181
	9.2 Technology for Monitoring the State of the Heart with the Use of the Spectral Characteristics of Heart Sounds	190
	9.3 Correlation System for Monitoring the Beginning of the Latent Period of Vascular Pathology in Human Body	194
	References	197
Index		199

Introduction

Varieties and stages of the initiation and development of defects preceding accidents at various facilities are analyzed. It is shown that the registration of the beginning of the latent period of transition of objects into an emergency condition based on the results of traditional technologies for analyzing measuring information in control systems is sometimes belated due to the impossibility of analyzing the noise correlated with the useful signal. This sometimes leads to disastrous accidents. Algorithms and technologies for calculating the estimates of correlation functions, spectral characteristics, and other noise characteristics are proposed. These algorithms allow forming corresponding sets of informative attributes for the control of the beginning of the latent period and development dynamics of accidents. The structural principles of various noise control intelligent systems, the possibilities of their integration into oil and gas production facilities, drilling rigs, offshore fixed platforms, compressor stations, in transport, aviation, power engineering, construction, seismology, and medicine are illustrated with examples. It is also shown that the use of the noise technologies can improve the accuracy of the results of traditional methods for analyzing noisy random signals.

Chapter 1
Specifics and Difficulties of Control of the Beginning of Accident Initiation and Development Dynamics

Abstract Types and stages of initiation and development of defects caused by fatigue, corrosion, contamination, overheating, overload, scuffing, wear, etc., and preceding accidents of technical facilities are considered. It is shown that they manifest themselves in the form of noise in the signals received at the outputs of the sensors and introduce an additional error in the results of processing by traditional analysis technologies in control systems. However, despite the difference in the process of defect initiation and development for different facilities, their common characteristic is that the beginning of the latent period of changes in their state manifests itself in the signals in the form of noises containing valuable diagnostic information. At the same time, development dynamics of an accident in its latent period affects the duration of the correlation time and the cross-correlation function between the useful signal and the noise. The chapter also considers the specifics of the data support to ensure solving the problem of control of the beginning of initiation and development dynamics of accidents. Models of signals received at the outputs of sensors in the latent period of accidents of control objects are analyzed.

1.1 Causes and Specifics of Accident Initiation

The causes of the initiation of defects preceding accidents in various objects, both in living organisms and in engineering, are the subject of research of the relevant fields of science. However, they have much in common in terms of obtaining information, methods of their analysis, control, and diagnostics. In many cases, the outputs of the sensors installed on the relevant objects produce signals that reflect their technical condition or the state of their functioning. The same information technologies are used for the analysis of these signals as an information carrier in various fields for completely different objects. They are implemented in the same modern computers. Therefore, from the point of view of a measuring and information systems and information technology specialist, solving the problem of controlling their state is not vastly different, despite the wide range of specifics of every object.

However, for every object, the process of defect initiation and development to a pronounced emergency state has significant distinctive features related to physical, biological, mechanical, chemical, and other properties of objects [1–9]. This is also related to the functions that they perform, their operation modes, etc. For these and many other reasons, the process of defect initiation to a pronounced emergency state proceeds in different ways. In some objects, this happens quickly. In others, this process takes much more time. However, despite all these differences, they have in common that in that period, the information contained in the signals continuously changes due to the emergence of a defect in the form of a weak component. It stabilizes only after a pronounced emergency state. For this reason, solving the problem of controlling the beginning of accident initiation and development dynamics differs from the diagnostics of the technical state of the object [10–17].

Here are some examples.

1. When a pinhole forms in an oil and gas pipeline, one first hears faint whistling, which then grows louder, then the pipe starts humming and finally snorting and gurgling [1, 2]. Therefore, the characteristics of the signal received at the output of the acoustic sensor change continuously this entire time.
2. When a microcrack forms in metal trusses and supports of a fixed offshore platform, one first hears faint squeaking, which then grows louder, then it reminds of a baby's crying, turns into metallic creaking, etc. [4].
3. At the first stage of weakening of the fixing point of the aircraft engine suspension spring, a high-frequency component also appears in vibration signals of the corresponding sensors. As this defect develops, the frequency gradually decreases, and this process continues until the suspension is completely detached [18].
4. Sticking of the drilling tool is not uncommon when drilling deep and superdeep wells under rock pressure [1, 2]. At the beginning of this process, weak high-frequency vibrations appear at the outputs of mechanical drilling speed sensors and torque sensors on the rotor of the drilling machine. In some cases, when they take on a pronounced form, preventing an accident becomes difficult. Unfortunately, sometimes the driller senses a disturbance in the smooth speed of the drill string when it rotates with obvious jitter. Starting from this moment, the driller undertakes technological operations to eliminate the sticking, not always succeeding.

It is known from [10] that when a crack appears in some parts of technical objects, intense acoustic radiation forms which continuously changes as it develops. Examples of such cracks emerging "in sensitive places" (for instance, near rivets) are cracks in the bodies of vehicles, including airliners, or microcracks in equipment (tanks) operating under high pressure.

It is obvious from the above examples that when developing a technology for detecting the effect of the beginning of initiation of various defects in the signals obtained from the relevant sensors, it is necessary to take into account that their spectra and other characteristics change continuously.

1.2 Types and Development Stages of Defects Preceding Accidents in Technical Facilities

The process of the initiation of defects preceding accidents during the operation of equipment can be classified based on the nature of the destructive effect (chemical, thermal, and mechanical) and the type of destruction. Defects in the components of equipment are mostly caused by fatigue, corrosion, pollution, overheating of overload, scuffing, wear, etc. Let us examine the most common variants of the initiation of these defects in detail [1–5, 18–64].

The initiation of a defect in the form of a crack as a result of metal fatigue is associated with the effects of cyclic loads. Fatigue point is a property of the material, but the degree of crack growth depends on many factors and is determined by the operating conditions [1, 2, 10].

As a rule, the process of crack development starts with the formation of microcracks, "coarsening" of the surface, grain boundary cracking, cracking around solid inclusions and is accompanied with further infiltration deep into the material. Sometimes, a microcrack turns into a macrocrack and spreads in the metal rather quickly.

Detection of the process of initiation and development of a crack in the early stages is of enormous importance for many modern critical facilities (turbines, offshore platforms, aircrafts, tankers, etc.). Even after they have been detected, for many objects, it is advisable to establish regular control of the detected cracks, since development of a crack can be indicative of the remaining life of the component before it is due for replacement. The relative rate of growth of fatigue cracks depends to a large extent on the quality of the material [1, 2, 10].

The causes of the initiation of defects in the form of a crack can be very different [1–5, 10].

Defects at a static load in the form of a crack emerge when a single load is applied, causing a stress exceeding the endurance limit of the material.

Defects from strain cause local deformation or "neck formation"; the crack surface is formed by separation planes inclined at an angle of about 45° to the load.

Defects from compression occur in two main forms: compression of the beam and bending (buckling). Compression of the beam occurs in short heavy parts separated by inclined planes in the same way as in strain. The only difference is that when the two halves of the crack are separated, there is friction.

Defect from bending occurs in long parts and causes a typical bending change in shape. The moment of bending applied to the material is resisted by the tensile and compressive stress of the material itself. As a result, the destruction of the material in this case is similar to the formation of cracks caused by tension from the outside of the bend and compression from the inside of the bend.

Bending is characteristic for metal sheets and occurs in such a way that the direction of the crests or troughs of the bending wave overlaps in the diagonal of the shifted plane.

In case of a defect from shearing, the two halves of the crack slide one over the other, the surface is subjected to friction, as a result of which the crack is smoothed or the surface is scuffed. The direction of the scuff shows the direction of application of the shearing force.

In the case of a torsional defect, the two halves of the destroyed metal sample retain some residual bending. The crack surface often is the same shape as in strain and is inclined at twisting angle.

Defect from the impact force breaks the lower part of the component into fragments when the middle part deforms and forms a void.

Wear is also one of the most common causes of defects. The process of initiation of defects in the form of wear of mating surfaces that are in relative motion to each other can be considered as one of the main causes of reduction of equipment service life.

Analyzing the nature of the change in wear rate, three distinct phases of wear of the component during its operation can be distinguished [1–5]. In the first phase, the breaking-in of components occurs, i.e., the micro- and macrostructure of surfaces change. Wear in the second phase is called the normal period. In many cases, a linear relationship can be assumed between the amount of wear and time. The main factors affecting the amount of wear in this phase are the specific pressure and the relative speed of movement of rubbing parts. With abrasive wear, the latter is directly proportional to the specific pressure on the friction surfaces and slide paths. The third phase, the emergency wear, is the result of quantitative changes in the structure of the surface of the mating parts, the process developing at a catastrophic pace.

Fatigue (pitting) wear usually occurs in rolling bearings and is caused by fatigue of the surface layer. In those cases when there is a relative slip of surfaces, fatigue wear is possible due to micro-roughness.

Molecular wear is characterized by the development of local metal compounds and the rejection of the formed particles from rubbing surfaces. This type of wear occurs at high pressure and, as a rule, develops quickly.

Corrosive wear occurs in an aggressive (oxidizing) medium. A cyclic load destroys the oxide (protective) film and reveals a fresh sublayer of metal, which in the presence of oxygen is oxidized, the resulting film disintegrates again, and the process is repeated.

Cavitation wear (cavitation erosion) is caused by local hydraulic impacts of fluid in the cavitation zone. If the component operates in a stream of hot gases, then the surface softens and oxidizes, and the split-off metal particles (gas erosion) are carried along with the gas flow.

One of dangerous defect types is caused by scuffing. When the surfaces of mating parts come into contact, a pressure can occur that is able to break the lubricating film and expose the mating surfaces, which creates conditions for welding of local areas. The damage to the surface that results in local adhesions between rubbing surfaces is called scuffing. This process is caused by a sharp increase in the coefficient of friction between rubbing surfaces.

1.2 Types and Development Stages of Defects Preceding Accidents ...

Scuffing is characteristic mainly for gear systems and pistons in internal combustion engines. It is due to the disturbance of the lubrication regime, abrupt heating, and destruction of the surface layer accompanied by intermolecular welding of the metal of two surfaces.

The initiation of a defect in the form of corrosion damage is the result of electrochemical and chemical processes occurring on the surface of a metal in a corrosive medium [1, 2, 10]. The nature of progress of corrosion in equipment is mainly determined by the operating conditions. Stress corrosion, corrosion fatigue of metal, and cavitation erosion are commonly distinguished.

Stress corrosion is the cause of metal cracking under constant stress in a corrosive medium. Cracks in this case are located normally to tensile stresses, have uneven rough edges, and can be intercrystalline or transcrystalline depending on the material.

Corrosion fatigue is characteristic for parts that operate in a corrosive environment and are exposed to tensile cyclic loads. The cracks that form in this case spread deeper into the material perpendicular to the tensile stresses and are mainly transcrystalline in nature and are usually filled with corrosion products.

Cavitation erosion processes usually occur in hydraulic mechanisms, such as ship propellers and rudders, cylinder liners of engines, sleeve bearings, fuel injection systems, pumps, and hydraulic systems.

The destruction of the material under cavitation is caused by the simultaneous mechanical effect of closing fluid bubbles and electrochemical corrosion.

In some cases, the cause of defect initiation is corrosion in a sulfuric medium. Most types of fuels contain sulfur impurities, which after combustion combine with condensed water vapor to form sulfuric acid in the form of sulfurous mist. For instance, corrosion in a sulfurous medium of guide vanes of marine gas turbines is described in [10].

Carbon formation is another most common causes of the initiation of various defects. In diesel engines, carbon deposits cause the coking up of injection nozzles, loss of mobility of piston rings, and other irregularities. In gas turbines, carbon deposits distort the front and structure of the flame, and cause warping of the combustion chambers, wear of turbine blades due to erosion and corrosion, etc.

Procedural violations in the process of assembly also sometimes cause the initiation of defects [1, 2, 10].

In conclusion, it should be noted that the control of the initiation of defects that precede accidents requires first of all determining the degree of threat those defects represent, with allowance for the dynamics of their change in time, and assessing possible results of their development under critical conditions. For this purpose, it is expedient to divide this process into three distinctly different stages: (a) control of the beginning of defect initiation, (b) identification of the defect and its stage, (c) determining and controlling the defect development dynamics.

1.3 Sensors and Specifics of Information Support for the Control of the Beginning of Accident Initiation and Development

By the control of the beginning of the latent period of accidents, we understand early detection of defect initiation in the initial stage of accidents, when no negative consequences for the reliability or operability of the equipment manifest themselves yet.

It is known that objects usually break down due to various defects [1–5, 10]. In some cases, those defects lead to catastrophic consequences. To prevent them, it is necessary to control the initiation of defects that precede such accidents. Solving the problem of controlling the beginning of a defect that leads to a violation of the integrity and operability of structures requires, first of all, creating appropriate technologies and software for analyzing the signals received at the outputs of the respective sensors. In doing that, it is essential to be able to continuously receive the necessary information for controlling the beginning of initiation of all kinds of defects. To this end, based on the statistics of the most dangerous accidents, it is necessary to determine the type of sensors and the location of their installation ("vulnerable spots"), ensuring that sufficient information is received from the object, making early detection of the defect initiation sufficiently reliable [1–5, 10].

In the control of the beginning of accident initiation, those sensors in which the signals received at the outputs reflect the beginning of the initiation of the most widespread defects have a great information potential. Technological parameters such as temperature, pressure, vibration, acoustic and thermal radiation, etc. contain sufficient information to control the initiation of relevant defects. For instance, for many technical objects, such types of information include the spectrum of vibration of structural elements, the spectrum of acoustic vibrations, and other parameters that characterize the operation of the system. Moreover, not only the values of these parameters at a given time are important but also their changing in time. This is due to the fact that during operation, equipment elements get displacements that change in time (vibrational displacements). Vibrational displacements can be caused by cyclic processes during the operation of equipment (rotation of rotors, periodic loads, etc.), natural vibrations of structural elements, etc. In general, each design point has a spatial displacement, which is the geometric sum of three displacement components. At any given time, vibrational displacement can be represented as a superimposition of elementary harmonic oscillations with different frequency and amplitude. For instance, internal combustion engines, pumps, hydraulic valves, elements of ball bearings, etc. are typical systems for which vibration during operation is typical.

Equipment for batch operation generates a vibration that repeats after a certain time interval. Temporary realization of this vibration can be successfully used to control a defect at the beginning of its initiation.

Induction and piezometric sensors are used as vibration sensors. Modern vibration sensors have high vibration resistance and heat resistance (up to 500 °C) and are attached to the part with a flange or screwed into a threaded hole. It is very important

1.3 Sensors and Specifics of Information Support for the Control ...

to choose the location of the axis direction of the location of vibration sensors. This is a critical operation in the vibration control of equipment. For instance, bearings are the best place to measure equipment vibration, since it is in these places that the main dynamic loads and forces are applied. The vibration sensor must be mounted on the housing of as many bearings of the equipment as possible. If it is impossible to implement, measurements should be taken at the location closest to the bearing, with the smallest possible impedance between the bearing and that point.

The reliability of accident control also depends on the quality and amount of the measurement information that can be obtained during operation of the object. Therefore, the choice on the number of sensors is very important.

As an example, it can be shown that in a system controlling the initiation and development of accidents on deepwater offshore platforms, the reliability of the results depends on the number of vibration sensors installed in the main sections and the most vulnerable "informative" structures of platforms. Naturally, as their number increases, the control system becomes more complicated, and its decrease leads to the probability that some faults will be detected with a delay. Therefore, the issue of optimizing the number of these sensors is also of great importance. But a preliminary analysis shows that the number of sensors should be at least 15–20. This issue is described in detail in [4].

In general, the process of selecting a sensor is determined by the conditions of its application. Vibration control systems most commonly use sensors of [1–5, 10]:

- vibration displacement;
- vibration velocity;
- vibration acceleration.

Sensors of the first group characterize control object's position, sensors of the second group—the rate of change of its displacement with respect to time, and sensors of the third group—the rate of change of velocity. These three parameters that characterize vibration are interrelated, and, controlling, for example, vibration acceleration through single or double integration, it is easy to calculate the remaining two parameters.

The presence of the three types of sensors is due to the need to control vibration at facilities with different frequency characteristics. Vibration displacement sensors have proved themselves in the low-frequency region, vibration velocity sensors are usually used for medium-frequency facilities, and vibration acceleration sensors for high-frequency ones.

For instance, based on the results of our analysis of possible applications of vibration sensors to control the technical condition of fixed offshore platforms, Bean Device AX-3D sensors have been found most suitable. They can be easily installed in the most sensitive spots of the platform structure [4]. The measurement information from the sensors is collected via Wi Fi using the Bean CetanWay Controller. The range of Wi-Fi signals from the Bean Device AX-3D sensor is up to 650 meters, which is sufficient, considering the size of offshore platforms.

Acoustic signals also carry information about the beginning of defect initiation. They have enormous information potential for detecting the beginning of the process of defect initiation in technical and biological objects [6]. For instance, in technical objects, an acoustic signal is caused by the movement and vibrations of parts of the machine and the impact of the working process on the surrounding atmosphere. In some machines, e.g., in aircraft engines, a powerful source of acoustic vibrations (noise) is the jet exhaust from the propelling nozzle, acoustic radiation from the compressor blades, etc. At the same time, when a crack appears, intense high-frequency acoustic radiation arises. The extraction of this high-frequency spectrum from the noisy acoustic signal can be used for early detection of cracks [1–5]. In a broad sense of the word, an acoustic signal is a noise from a mechanical vibration transmitted through the air and containing a combination of different frequencies with different pressure levels. It is often a stochastic process whose amplitudes and frequencies are random. The composition of the spectrum and its amplitude–frequency response can contain important information about the technical state of the machine. Experienced mechanics are known to be able to determine the nature of malfunction in engines, turbines, etc., "by ear". Thus, it is only natural that by measuring acoustic oscillations and performing their spectral and correlation analysis, we can find the relevant informative attributes for the control. To this end, microphones based on electrical or piezoelectric effects with a frequency range of 5–100 kHz (the frequency of "audible" sound is 20 kHz) are used. The main difficulty in using vibroacoustic methods is the isolation of a useful signal from the background of noise. In practice, filters are often used to detect the attributes that carry diagnostic information [1, 2, 10–17].

In the control of the beginning of the latent period of accidents, it is natural that in addition to the above, signals received at the outputs of pressure, temperature, force, displacement, acceleration, and other sensors are also analyzed.

In many objects, noisy signals obtained at the outputs of sensors carry information about accident initiation come in the form of a random function. This is due to the fact that during accident initiation, noise is the result of imposition of a large number of various dynamic impacts emerging in equipment elements (natural and forced oscillations, collisions, effects of the working and external environment, etc.). Therefore, a noisy signal of chaotic, random nature contains sufficient information about the technical condition of the object. For instance, vibration signals of engines contain a large number of different noises. They make it difficult to detect the beginning of defect initiation using traditional signal analysis technologies. At the same time, in some cases, it is noise that contains diagnostic information about the beginning of accident initiation. Therefore, it is necessary to develop technologies that allow us to determine the informative attributes of both useful signals and noise [1–9, 18–64].

Thus, according to the above, in order to ensure control of a defect at the beginning of its initiation, it is first and foremost necessary to choose the type, quantity and location of the relevant sensors that ensure the object's controllability. In addition, for the analysis of signals received from these sensors, it is advisable to use the information technologies that make it possible to obtain relevant informative attributes of both the sum signal $g(i\Delta t)$ and the noise $\varepsilon(i\Delta t)$. The corresponding current and reference sets of informative attributes formed on their basis will be the foundation of information support for the problem of controlling the beginning of accident initiation and development dynamics.

1.3 Sensors and Specifics of Information Support for the Control ...

Traditional technologies do not allow extracting sufficient diagnostic information about the beginning of the latent period of accident initiation. And this affects the reliability of the results of control, which sometimes leads to errors in which accidents with catastrophic consequences become inevitable. Therefore, in order to increase the operational reliability of an object by early detection of the beginning of accident initiation and organization of timely maintenance, it is necessary to create new, more efficient technologies for analyzing noisy signals.

1.4 Models of Signals Received at the Sensor Outputs in the Latent Period of Accidents

As mentioned in Sect. 1.3, the information support for solving the problem of control of accident initiation and development dynamics suggests that the relevant measuring information is available. For this purpose, sensors are used that convert nonelectric quantities of temperature, pressure, mechanical displacement, vibration, etc., to electrical signals. In the general case, models of these signals, when the object is in the normal state, can be represented in the form

$$g(t) = X(t) + b(t) + c(t) + e(t)$$

Here, the quantity $X(t)$ is the useful signal corresponding to the measured technological parameter. The function $b(t)$ is determined by slow changes in operating conditions or characteristics of the equipment, raw material properties, daily load changes, etc. The function $c(t)$ is formed under the influence of various external factors (pressure, temperature, humidity, magnetic field, etc.). The function $e(t)$ is a measurement noise that occurs in sensitive elements, communication channels, measuring instruments, and converters. Thus, when the object is in the normal state, the technological parameter is the noisy signal $g(t)$, which consists of the useful signal $X(t)$ and the sum noises $b(t)$, $c(t)$, $e(t)$.

The value of $b(t)$ changes slowly and therefore has practically no effect on the result of signal analysis. On modern sensors, the effects of $c(t)$ on the measurement result are minimized due to screening, sealing, and other engineering solutions. Due to the use of the modern microelectronics, components in analog-to-digital converters, the effect of $e(t)$ on the measurement result is minimized as well.

Despite the above, band-pass filters are used at the output of the sensors in many information systems to eliminate the effects of $c(t)$ and $e(t)$ on the result of signal analysis. Obviously, effective filtering requires a coincidence of the spectrum of the sum noise

$$\varepsilon_1(t) = b(t) + c(t) + e(t)$$

and the filter band.

Otherwise, filtering can lead to a distortion of the spectral composition of the useful signal [1–9, 53–59]. Therefore, in practice, on the basis of a priori information about the possible spectral composition of $\varepsilon_1(t)$, a filter band is carefully chosen for each signal, which is corrected experimentally. In this case, the filtration operation justifies itself.

Another possible way to eliminate the effects of the noise $\varepsilon_1(t)$ on the result of signal analysis is the robust technology of correlation and spectral analysis [39–52]. According to this technology, the robustness value is determined in parallel in the process of analyzing noisy signals. Its value is used to correct the processing results from the effect of the noise. However, the algorithms and signal analysis programs become more complicated.

Thus, in real monitoring and control systems of different purpose, the noisy signals $g(t)$ obtained at the sensor outputs are the sum of the useful signal $X(t)$ and noise $\varepsilon_1(t)$, i.e.,

$$g(t) = X(t) + \varepsilon_1(t).$$

In this case, the known classical conditions are fulfilled for the normal state of the object for the centered noisy signals $g(t) = X(t) + \varepsilon_1(t)$ obtained at the outputs of the corresponding sensors, i.e.,

$$\left.\begin{aligned} M[X(t)X(t)] \neq 0, \ M[\varepsilon_1(t)X(t)] = 0 \\ M[X(t)\varepsilon_1(t)] = 0, \ M[\varepsilon_1(t)\varepsilon_1(t)] \neq 0 \end{aligned}\right\} \quad (1.1)$$

As a result, the formula for calculating the variance D_{gg} of the signal $g(t)$ takes the form

$$\begin{aligned} D_{gg} &= M[g(t)g(t)] = M[(X(t) + \varepsilon_1(t))(X(t) + \varepsilon_1(t))] \\ &= M[X(t)X(t) + X(t)\varepsilon_1(t) + \varepsilon_1(t)X(t) + \varepsilon_1(t)\varepsilon_1(t)] \\ &= M[X(t)X(t) + \varepsilon_1(t)\varepsilon_1(t)] \end{aligned}$$

It follows that

$$D_{gg} = M[X(t)X(t) + \varepsilon_1(t)\varepsilon_1(t)] = D_{XX}(0) + D_\varepsilon,$$

where

$$D_{\varepsilon_1} = M[\varepsilon_1(t)\varepsilon_1(t)] = M[\varepsilon(t)\varepsilon(t)] = D_\varepsilon.$$

At the beginning of the latent period of initiation and development of accidents, the noise $\varepsilon_2(t)$ correlated with the useful signal $X(t)$ emerges in the signals $g(t)$ received from the corresponding sensors in control objects. Thus, the period T_0 of the normal state of the object's operation ends, the period T_1 of latent change of its technical condition starts, and conditions (1.1) are violated. This happens during

1.4 Models of Signals Received at the Sensor Outputs in the Latent ...

the operation of any object when there are inevitable defects from fatigue, wear, corrosion, vibration, etc. In this case, the noise $\varepsilon_2(t)$ forms, the spectrum of which varies continuously from the moment of its nucleation to the moment it becomes pronounced. The noise $\varepsilon_2(t)$ is divided into mechanical, acoustic, electromagnetic, etc., depending on the source of origin. Depending on the type of signal and noise, they are divided into pulsed and continuous. Depending on the location of the defect, they are divided into internal and external. As will be shown below, starting with the moment of the emergence of the noise $\varepsilon_2(t)$, an additional error arises in the estimate of D_{gg}.

At the same time, from the beginning of the process of defect initiation to the moment it becomes pronounced, in addition to the spectrum, the power of the noise $\varepsilon_2(t)$ changes as well. Thus, the process of accident initiation is reflected on the signal $g(t)$ at the output of the sensor in the form of the noise $\varepsilon_2(t)$. Therefore, starting from the moment of initiation and development of the latent period of accidents, the model of the signal $g(t)$ at the sensor output can be represented in the form

$$g(t) = X(t) + \varepsilon_1(t) + \varepsilon_2(t). \tag{1.2}$$

At the same time, the presence of a correlation between the useful signal $X(t)$ and the sum noise

$$\varepsilon(t) = \varepsilon_1(t) + \varepsilon_2(t),$$

leads to the following inequalities:

$$\begin{cases} M[X(t)X(t)] \neq 0, \\ M[X(t)\varepsilon_2(t)] \neq 0, \\ M[\varepsilon_1(t)\varepsilon_1(t)] \neq 0, \\ \varepsilon(t)\varepsilon(t) \neq 0, \\ \varepsilon(t)X(t) \neq 0. \end{cases}$$

and equalities

$$\begin{cases} M[\varepsilon_1(t)X(t)] = 0, \\ \varepsilon_1(t)\varepsilon_2(t) = 0. \end{cases}$$

Consequently, we have

$$\begin{aligned} D_{gg} &= M\{[X(t) + \varepsilon_1(t) + \varepsilon_2(t)][X(t) + \varepsilon_1(t) + \varepsilon_2(t)]\} \\ &= M[X(t)X(t) + X(t)\varepsilon_1(t) + X(t)\varepsilon_2(t) + \varepsilon_1(t)X(t) + \varepsilon_1(t)\varepsilon_1(t) \\ &\quad + \varepsilon_1(t)\varepsilon_2(t) + \varepsilon_2(t)X(t) + \varepsilon_2(t)\varepsilon_1(t) + \varepsilon_2(t)\varepsilon_2(t)] \\ &= M[X(t)X(t) + \varepsilon_2(t)X(t) + X(t)\varepsilon_2(t) + \varepsilon_1(t)\varepsilon_1(t) + \varepsilon_2(t)\varepsilon_2(t)] \\ &= R_{XX}(t) + 2R_{X\varepsilon_2}(t) + D_{\varepsilon_1\varepsilon_1} + D_{\varepsilon_2\varepsilon_2} \end{aligned} \tag{1.3}$$

where

$$D_\varepsilon = M[\varepsilon_2(t)X(t) + X(t)\varepsilon_2(t) + \varepsilon_1(t)\varepsilon_1(t) + \varepsilon_2(t)\varepsilon_2(t)]$$
$$= 2R_{X\varepsilon_2}(0) + D_{\varepsilon_1\varepsilon_1} + D_{\varepsilon_2\varepsilon_2} = 2R_{X\varepsilon} + D_{\varepsilon\varepsilon}$$
$$D_{\varepsilon\varepsilon} = D_{\varepsilon_1} + D_{\varepsilon_2} \tag{1.4}$$

It is obvious from these formulas that in the latent period of accident initiation and development, estimates of the statistical characteristics of the signals $g(t)$ will be calculated with a certain error if traditional technologies are used. For this reason, the detection of the initial stage of accident initiation in the period T_1 becomes difficult. After the period T_1 ends, the period T_2 starts, which is when the emergency state becomes pronounced. When obtained by traditional technologies, the results indicating the beginning of the transition of an object to an emergency state in some cases turn out to be belated.

It is obvious from expressions (1.3) and (1.4) that the estimates of $R_{X\varepsilon_2}(t)$ and $D_{\varepsilon_1\varepsilon_2}$ reflect the effects of defect initiation on $g(t)$ and carry information on the beginning and dynamics of accident development. From this, it follows that in order to successfully solve the problem of controlling the beginning of accident initiation, it is necessary to extract the information contained in the noise $\varepsilon_2(i\Delta t)$. Therefore, in order to increase the reliability and validity of the results of control of the latent period of accident initiation and development, it is advisable to create technologies that allow maximum extraction of information contained in the noise $\varepsilon_2(t)$. These issues are discussed in detail in Chaps. 2 and 3.

1.5 Difficulties in Controlling the Beginning of the Latent Period of Accidents with the Use of Traditional Technologies

Numerous accidents with catastrophic consequences at technical facilities such as thermal and nuclear power plants, large-capacity petrochemical complexes, deepwater fixed offshore platforms and hydraulic structures, as well as airliner accidents, errors in earthquake forecasting, difficulties in diagnosing diseases in early stages, etc. used to be associated with the unreliability of components and the inaccuracy of measuring instruments. However, even after a multiple increase in reliability and accuracy of equipment, the probability of accidents continues to be significant.

Our analysis shows that in many cases, the main reason for information systems making inadequate solutions is due to inaccuracies in the results of the processing of measurement information and the lack of the possibility to ensure the robustness of the sought estimates. The formation of adequate solutions even in the most sophisticated and expensive control and management systems is impossible without the extraction of maximum information contained in noisy signals received at the outputs of corresponding sensors [53–59].

1.5 Difficulties in Controlling the Beginning of the Latent Period …

At the same time, with the use of traditional technologies, it is possible to obtain more or less acceptable estimates of the sought-for characteristics only if the analyzed signals are stationary, obey the normal distribution law, the correlation between the noise and the useful signal is zero, and the noise is white noise [10–17]. However, even in this case, the errors of the found estimates depend on the correctness of the sampling interval, time of implementation, etc. For this reason, even when these conditions are fulfilled, the adequacy of describing many analyzed processes by means of traditional technology is unsatisfactory, and when solving numerous important problems, erroneous results are obtained [1, 2]. This is due to the fact that when developing relevant algorithms, the specifics of the formation of real signals are insufficiently taken into account. At the same time, in most cases, these conditions are not met at all. The conducted studies showed that the signals received at the outputs of many modern sensors contain a great information potential. They are quite sufficient for forecasting accidents at thermal and nuclear power plants, for predicting accidents during drilling, for preflight control of the technical condition of aircrafts, for forecasting earthquakes and other various natural disasters, for early diagnosis of diseases, for increasing the efficiency of geophysical prospecting of minerals, etc. [1–9, 18–64]. In this regard, it has long become necessary for the full use of the colossal information potential of signals received at the outputs of modern sensors to revise the traditional algorithms and create new technologies that would ensure the adequacy of the results of tasks in information systems both under the anticipated classical conditions and when they are not fulfilled. To do this, we need technologies that make it possible to extract as much information as possible from the signals. One of the possible solutions to this problem is to create technologies that can extract information from the noise of noisy signals. In general, such a technology would contribute to the further development of the theory of correlation analysis, the theory of spectral analysis, the theory of pattern recognition, the theory of random processes, etc. At the same time, it will be possible to expand the boundaries of solving the most important problems in physics, biology, energy, geology, petrochemistry, aviation, etc.

1.6 Factors Affecting the Adequacy of the Control of Accident Initiation by Correlation Analysis Methods

Let us consider the factors that complicate solving the problems of controlling accident initiation using correlation analysis methods, for the case when the estimates of the correlation functions are calculated using the measurement information $g(t) = X(t) + \varepsilon(t)$, $\eta(t) = Y(t) + \varphi(t)$ that consists of the useful signals $X(t)$, $Y(t)$ and the noises $\varepsilon(t)$, $\varphi(t)$, which obey the normal distribution law with the mathematical expectations $m_x \approx 0$, $m_y \approx 0$, $m_\varepsilon \approx 0$, $m_\varphi \approx 0$, respectively.

It is known that the formula for calculating the estimates of the auto- and cross-correlation functions $R_{gg}(\tau)$ and $R_{g\eta}(\tau)$ between the signals $g(t)$ and $g(t)$ and between $g_i(t)$ and $\eta(t)$, respectively, is

$$R_{gg}(\tau) = \frac{1}{T}\int_0^T g(t)g(t+\tau)dt$$

$$= \frac{1}{T}\int_0^T [X(t)+\varepsilon(t)][X(t+\tau)+\varepsilon(t+\tau)]dt$$

$$= \frac{1}{T}\int_0^T [X(t)X_i(t+\tau)+X(t)\varepsilon_i(t+\tau)+\varepsilon(t)X_i(t+\tau)$$

$$+\varepsilon(t)\varepsilon_i(t+\tau)]dt \tag{1.5}$$

$$R_{g\eta}(\tau) = \frac{1}{T}\int_0^T g(t)\eta(t+\tau)dt = \frac{1}{T}\int_0^T [X_i(t)+\varepsilon_i(t)][Y(t+\tau)$$

$$+\varphi(t+\tau)]dt$$

$$\frac{1}{T}\int_0^T [X(t)Y(t+\tau)+X(t)\varphi(t+\tau)+\varepsilon_i(t)Y(t+\tau)$$

$$+\varepsilon(t)\varphi(t+\tau)]dt \tag{1.6}$$

Considering that for real signals obtained at the sensor outputs, the equalities

$$\frac{1}{T}\int_0^T \varepsilon(t)\varepsilon(t+\tau)dt \approx 0, \quad \frac{1}{T}\int_0^T \varepsilon(t)\varphi(t+\tau)dt \approx 0, \tag{1.7}$$

are true, we get the following expressions for the errors of the cross-correlation functions $R_{gg}(\tau)$ and $R_{g\eta}(\tau)$:

$$\Lambda_{xy}(\tau) = \frac{1}{T}\int_0^T X[x(t)\varphi(t+\tau)+\varepsilon(t)Y(t+\tau)]dt \approx 0 \tag{1.8}$$

The value of the error $\Lambda_{xx}(\tau)$ of the estimates of the autocorrelation functions $R_{gg}(\tau)$ is calculated from the expression

1.6 Factors Affecting the Adequacy of the Control of Accident ...

$$\Lambda_{xx}(\tau) = \frac{1}{T}\int_0^T [X(t)\varepsilon(t+\tau) + \varepsilon(t)X(t+\tau) + \varepsilon(t)\varepsilon(t+\tau)]dt. \quad (1.9)$$

However, taking into account that the values of $\varepsilon(t)$ and $\varepsilon(t+\tau)$ at $\tau \neq 0$ do not correlate with each other, i.e.,

$$\frac{1}{T}\int_0^T \varepsilon(t)\varepsilon(t+\tau)dt = 0 \text{ when } \tau \neq 0, \quad (1.10)$$

then equality (1.9) turns out to be valid only for the values $\tau = 0$. For all other values $\tau \neq 0$, the following equality will hold:

$$\Lambda_{x_i x_i}(\tau) = \frac{1}{T}\int_0^T [X_i(t)\varepsilon_i(t+\tau) + \varepsilon_i(t)X_i(t+\tau)]dt. \quad (1.11)$$

When $\tau = 0$, the expression for the error $\Lambda_{x_i x_i}(\tau)$ assumes the following form:

$$\Lambda_{x_i x_i}(\tau) = \frac{2}{T}\int_0^T X_i(t)\varepsilon_i(t)dt + \frac{1}{T}\int_0^T \varepsilon_i^2(t)dt. \quad (1.12)$$

Taking into account that the mean square value of the noise is equal to the variance D_ε of the noise $\varepsilon_i(t)$, i.e.,

$$\frac{1}{T}\int_0^T \varepsilon^2(t)dt = D_\varepsilon, \quad (1.13)$$

at $\tau = 0$, we get

$$\Lambda_{x_i x_i}(\tau) = \frac{2}{T}\int_0^T X_i(t)\varepsilon_i(t)dt + D_\varepsilon = \Lambda'_{xx}(\tau) + D_{\varepsilon_i}, \quad (1.14)$$

where

$$\Lambda'_{x_i x_i}(\tau) = \frac{2}{T}\int_0^T X(t)\varepsilon(t)dt.$$

It is obvious from these expressions that the estimate of $R_{gg}(0)$ has an error equal to the value of $\Lambda_{xx}(0)$, and the estimate of $R_{gg}(\tau)$, in addition to the values of the error $\Lambda_{xx}(0)$, also contains an additional error equal to the value of the variance D_ε of the noise $\varepsilon_i(t)$. The estimate of $R_{gn}(0)$ does not have an error.

Thus, it is obvious from expressions (1.5)–(1.14) that even in the normal mode of operation of the object and under the mentioned classical conditions, the inadequacy of the results of solving the problem of controlling the beginning of accident initiation is caused by the presence of the errors $\Lambda_{xx}(0)$ in every estimate, as well as the variance D_ε of the noise $\varepsilon_i(t)$ in these estimates at $\tau = 0$.

Therefore, for the problem of control of accident initiation, it is impossible to get results that would be adequate to the situations, if the signal model is represented in the form of expressions (1.2). If we take into account that for many objects, even under normal operating conditions, when solving the problem under consideration, the classical conditions are not fulfilled at all, then it becomes clear why the use of traditional correlation analysis technologies in the latent period of the emergency state of the object does not yield adequate results.

Thus, to ensure the reliability of the results of controlling the beginning of accident initiation, we need to create more advanced digital correlation analysis technologies [39–48].

1.7 Factors Affecting the Adequacy of the Control of Accident Initiation by Spectral Analysis Methods

It is known that in spectral analysis, when the measurement information consists of the useful signal and the noise, the error of the sought-for estimates depends on the difference between the sum of the errors of the positive and negative products of the samples of the sum signal [1, 2, 49–52].

If the analyzed signal $X(t)$ does not contain the noise $\varepsilon(t)$, then it can be represented as a sum of harmonic functions (a sine and a cosine wave), the sum of the ordinates of which at any given time t gives the value of the original function [1, 2, 10–17]

$$X(t) = \frac{a_0}{2} + \sum_{i=1}^{\infty}(a_n \cos n\omega t + b_n \sin n\omega t), \quad (1.15)$$

where $\frac{a_0}{2}$ is the mean value of the function $X(t)$ over the period T; a_n and b_n are the amplitudes of the cosine wave and sine wave with the frequency $n\omega$.

The necessary accuracy of the description of the signal $X(t)$ by the sum of cosine and sine waves requires that the following inequality be true:

$$\sum_{i=1}^{\infty} \lambda_i^2 \leq S. \quad (1.16)$$

1.7 Factors Affecting the Adequacy of the Control of Accident ...

Here, λ_i^2 are the squared deviations between the sum of the right-hand side of equality (1.15) and the samples of the signal $X(t)$ at the sampling instant $t_0, t_1, \ldots, t_i, \ldots, t_m$ with the interval Δt; S is the allowable value of the mean square deviation.

In Formula (1.15), when the function $X(t)$ is expanded in a trigonometric Fourier series, ω is assumed to be equal to $\frac{2\pi}{T}$, and the coefficients a_n and b_n are calculated as follows:

$$a_n = \frac{2}{T} \int_0^T X(t) \cos n\omega t \, dt \text{ for } n = 1, 2, \ldots; \quad (1.17)$$

$$b_n = \frac{2}{T} \int_0^T X(t) \sin n\omega t \, dt \text{ for } n = 1, 2, \ldots. \quad (1.18)$$

Here, the first harmonic curve has a frequency of $\frac{2\pi}{T}$, and its period is the same as the period T of the function $X(t)$. The coefficients a_1 and b_1, a_2 and b_2, a_3 and $b_3, \ldots, a_n$ and b_n are the amplitudes of the cosine and sine waves obtained at $n = 1$, $n = 2$, $n = 3$, etc.

Theoretically, for the signals $X(t)$ with a bounded spectrum, if they do not contain the noise $\varepsilon(t)$, condition (1.16) for a given value of S is fulfilled. However, in reality, as mentioned above, the useful signal $X(t)$ is often accompanied by a certain noise $\varepsilon(t)$, i.e., is the sum $g(t) = X(t) + \varepsilon(t)$. For this reason, condition (1.16) is not always fulfilled. Nevertheless, for cases when the value of the noise varies within acceptable limits and obeys the normal distribution law, many of the most important problems can be successfully solved using algorithms (1.17) and (1.18). In practice, when analyzing the operation of linear elements and systems, the principle of superposition of signals is widely used, the essence of which is as follows. When presenting the input signal as the sum of a number of other signals, the output signal is determined as the sum of the output signals that would be received if each of the input signals acted separately. This is the simplest way to determine, by means of harmonic analysis of the response of a linear system or the linear part of a nonlinear system to an input signal of arbitrary shape. That is why in the control of defect initiation, it is particularly advisable to use methods and algorithms of spectral analysis. However, when the relevant classical conditions are not fulfilled and the size of the noise is appreciable, it is impossible to ensure the fulfillment of inequality (1.16). For this reason, it is difficult to solve the problems of controlling the beginning of defect initiation using spectral methods. Let us consider this issue in more detail.

For the case when the analyzed signal $g(t)$ consists of the useful random signal $X(t)$ and the noise $\varepsilon(t)$, i.e.,

$$g(t) = X(t) + \varepsilon(t),$$

Formula (1.17) assumes the following form:

$$a_n = \frac{2}{T}\int_0^T [X(t)+\varepsilon(t)]\cos n\omega t\,dt = \frac{2}{T}\int_0^T \{[X(t)\cos n\omega t\,dt] + [\varepsilon(t)\cos n\omega t\,dt]\}.$$

Obviously, that condition (1.15) can hold for the cases when

$$\sum_{i=1}^{N^+}\int_{t_i}^{t_{i+1}} \varepsilon(t)\cos n\omega t\,dt = \sum_{i=1}^{N^-}\int_{t_{i+1}}^{t_{i+2}} \varepsilon(t)\cos n\omega t\,dt,$$

where N^+, t_i, t_{i+1} are the number, the start and the end, respectively, of the positive half-cycles of $\cos n\omega t$ in the observation T; N^-, t_{i+1}, t_{i+2} are the number, the start and the end, respectively, of the negative half-cycles of $\cos n\omega t$ in the observation T.

In all other cases, when this equality is not fulfilled, we get the difference

$$\lambda_{a_n} = \sum_{i=1}^{N^+}\int_{t_i}^{t_{i+1}} \varepsilon(t)\cos n\omega t\,dt - \sum_{i=1}^{N^-}\int_{t_{i+1}}^{t_{i+2}} \varepsilon(t)\cos n\omega t\,dt \qquad (1.19)$$

which leads to an error in the estimate of the coefficient a_n. The same thing happens when determining the estimate of b_n. And, as we can see from expression (1.19), as the error $\varepsilon(t)$ grows, so does the difference λ_{a_n}. A deviation of the distribution law of the analyzed signal $X(t)$ and the noise $\varepsilon(t)$ also leads to an increase in the difference λ_{a_n}. For these reasons, in some cases, the errors in the estimates λ_{a_n} and λ_{b_n} can be commensurable with the sought-for coefficients a_n and b_n.

In this regard, for real objects, for which the abovementioned classical conditions are not fulfilled, reliable results of the control of the beginning of accident initiation are not always guaranteed.

To eliminate this shortcoming of spectral analysis technology, it is first of all necessary to develop algorithms and technologies that allow us to ensure the inequalities $S_n \gg \lambda_{a_n}$, $S_n \gg \lambda_{b_n}$ and condition (1.16) by eliminating the causes of the errors λ_{a_n} and λ_{b_n}. They need to satisfy the conditions of robustness, i.e., allow us to eliminate the relationship between λ_{a_n} and λ_{b_n} and t the variance D_ε of the noise $\varepsilon(t)$, the distribution law of the analyzed signal, the correlation coefficient between the useful signal $X(t)$ and the noise $\varepsilon(t)$, etc. It is also necessary to create a technology for the separate calculation of the estimates of the spectral characteristics of both the useful signal and the noise. A solution to these problems is discussed in detail in Chap. 4.

1.8 Effects of the Signal Filtration on the Result of the Control of the Beginning of Defect Initiation

It is known that one of the most common methods of eliminating the effects of the noise $\varepsilon(t)$ on the result of the analysis of the noisy signals $g(t) = X(t) + \varepsilon(t)$ is to filter the noise $\varepsilon(t)$ [1, 2, 15–22]. To perform the filtering of the noise of the noisy continuous signals received at the sensor outputs, we need to know the distribution of its mean power in frequency, to find its power spectral density, spectral width, position and values of the maximums of the power spectral density, boundary frequencies, etc. After obtaining these characteristics, it is also expedient to calculate the estimates of the power spectral densities of the noises of the measured signals. The calculation of these signal characteristics in spectral analysis is in most cases based on the Fourier transform of the signal. For noisy random signals, the method of calculating the power spectral density from the measured correlation function is often used in accordance with the Wiener–Khinchin theorem. The spectral power density $G_g(f)$ allows us to estimate the frequency properties of both the useful signal $X(t)$, and the noise $\varepsilon(t)$, since it is indicative of its intensity at different frequencies, i.e., average power per unit of frequency band. Signal functions are also used for this purpose, and there is a method that is based on the hardware application of orthogonal functions.

The hardware-based determination of the noise spectrum is based on applying the best known method of filtering by isolating narrow regions of the spectrum of the analyzed signal using an instrument with a selective amplitude–frequency response. It is assumed that if the noise spectrum is limited to frequencies $f_1 = f - \Delta f/2$ and $f_2 = f + \Delta f/2$, then the average power in the band Δf near the frequency f can be calculated from the expression

$$P_x(f, \Delta f) = 2 \int_{f-\Delta f/2}^{f+\Delta f/2} G_x(f) df.$$

If the frequency band Δf is finite but so narrow that the spectral power density $G_\varepsilon(f)$ can be assumed as constant in this band, then it can be calculated using the approximate expression

$$G_\varepsilon(f) \approx \frac{P_k(f, \Delta f)}{\Delta f}.$$

According to this formula, the filtering of the noise comes down to calculating the power spectral density by measuring its average power in the known narrow band Δf. In other words, to measure the power spectral density of $\varepsilon(t)$, we need to "cut out" a narrow band of the sum signal spectrum by means of a linear band-pass filter with the pass band Δf. Band-pass filter analyzers are used for this purpose. In this case, it is necessary to specify the noise spectrum and the filtering of the noise $\varepsilon(i\Delta t)$ from the sum signal $g(i\Delta t)$ can be represented as

$$X(t) \approx g(t) - \varepsilon(t) = \frac{a_0}{2} + \sum_{n=1}^{\infty} \left(a_{n_g} \cos n\omega t + b_{n_g} \sin n\omega t \right)$$
$$- \sum_{n=k}^{k+m} \left(a_{n_\varepsilon} \cos n_\varepsilon \omega t + b_{n_\varepsilon} \sin n_\varepsilon \omega t \right)$$

It is clear that such an ideal separation of harmonic curves of the noise within the range from $n = k$ to $k + m$ is impossible in reality. Naturally, in reality, there will be at least low-power spectra of the useful signal $g(t)$ within this range. At the same time, there will also be certain low-power spectra of the noise $\varepsilon(t)$ outside this range. For this reason, filtering with the use of band-pass filters gives good results in those rare ideal cases where noise spectra are within a given range. In all other cases, the filtration distorts certain spectra of both the noise and the useful signal. In view of the above, the low-power noise $\varepsilon_2(t)$, which occurs at the beginning of accident initiation, will be lost during filtering. For this reason, when filtering is applied, the beginning of defect initiation can be identified only at the time when it becomes pronounced. Therefore, in order to detect the beginning of defect initiation and development, it is necessary to develop digital technologies that eliminate the effects of noise on the result of processing, as well as to isolate and analyze noise and useful signals [53, 59]. These issues are discussed in Chaps. 3 and 4.

1.9 Effects of Traditional Methods of Sampling Interval Selection on the Adequacy of Control of the Beginning of Accident Initiation

It is known that, according to traditional methods, the sampling step Δt of signals at the sensor outputs is determined on the basis of its high-frequency low-power spectrum with the cutoff frequency f_c by condition

$$\Delta t \leq \frac{1}{2 f_c}.$$

With this way of determining the interval Δt, the estimates of the correlation and spectral characteristics of both the sum signal and the useful signal $X(t)$ are calculated with sufficient accuracy. However, in this case, the information carried by the high-frequency low-power spectra of the noise $\varepsilon(i \Delta t)$ of the sum signal is practically lost. This, in turn, leads to the loss of the information about the beginning of the initiation of defects that precede accidents. Therefore, it is necessary to take into account the effects of the noise $\varepsilon(i \Delta t)$ on the calculated estimates. To this end, we need to find the values of the samples of the sum signal, starting from the spectrum frequency of the noise $\varepsilon(i \Delta t)$, using the formula

1.9 Effects of Traditional Methods of Sampling Interval Selection ...

$$\Delta t \leq \frac{1}{2f_\varepsilon}.$$

Let us consider the possibilities of sampling signals with such frequency in more detail.

It is known that the speed of modern analog-to-digital converters (ADCs) and information systems makes it possible to convert the analyzed signals $g(t)$ with the oversampling frequency f_u:

$$f_u \gg f_s,$$

where f_s is the sampling frequency determined by traditional methods.

It is clear that with such oversampling of the signal $g(i\Delta t)$, many samples $g(i\Delta t)$ that are consecutive will be repeated and the frequency of the state change f_{q0} of the least significant bit of the analog-to-digital converter f_0 will be significantly smaller than the oversampling frequency, i.e.,

$f_u \gg f_{q0}.$

In this case, the equality

$$P[g(i\Delta t) \approx g((i+1)\Delta t)] \approx 1,$$

where P is the probability sign, will take place.

At the same time, it is intuitively clear that in this case, the frequency of the change of the least significant bit f_0 will largely depend on the high-frequency spectrum of $\varepsilon(t)$ of the signal $g(t)$, and the value of f_{q0} will essentially be the sampling frequency of the noise $\varepsilon(t)$ [64]. Therefore, to preserve the diagnostic information of the noise $\varepsilon(t)$, it is necessary to sample the sum signal $g(t)$ at the frequency Δt_ε. Then, the information on the beginning of defect initiation contained in the samples of the sum signal $g(t)$, will not be lost and it will be possible to control the beginning and the development dynamics of accidents in the periods T_1, T_2, T_3. Naturally, the technologies for analyzing the signals $g(t)$, in these time intervals must be adaptive, since the frequency characteristics of $\varepsilon(t)$ vary continuously in the periods T_1, T_2, T_3 with the high dynamics of accident development. It is also important to ensure the adaptability of the sampling of $g(t)$, because some of the extracted information contained in $\varepsilon(t)$ is lost both in case of insufficient and oversampling frequency of $g(t)$, [2]. The technology for ensuring the adaptability of the sampling step of $g(t)$, is discussed in more detail in several sections of the book.

References

1. Aliev T (2007) Digital noise monitoring of defect origin. Springer, Boston. https://doi.org/10.1007/978-0-387-71754-8
2. Aliev T (2003) Robust technology with analysis of interference in signal processing. Kluwer Academic/Plenum Publishers, New York. https://doi.org/10.1007/978-1-4615-0093-3
3. Aliev TA, Rzayev AH, Guluyev GA et al (2018) Robust technology and system for management of sucker rod pumping units in oil wells. Mech Syst Signal Process 99:47–56. https://doi.org/10.1016/j.ymssp.2017.06.010
4. Aliev TA, Alizada TA, Rzayeva NE et al (2017) Noise technologies and systems for monitoring the beginning of the latent period of accidents on fixed platforms. Mech Syst Signal Process 87:111–123. https://doi.org/10.1016/j.ymssp.2016.10.014
5. Aliev TA, Guluyev GA, Pashayev FH et al (2012) Noise monitoring technology for objects in transition to the emergency state. Mech Syst Sig Process 27:755–762. https://doi.org/10.1016/j.ymssp.2011.09.005
6. Aliev TA, Rzayeva NE, Sattarova UE (2017) Robust correlation technology for online monitoring of changes in the state of the heart by means of laptops and smartphones. Biomed Sig Process Control 31:44–51. https://doi.org/10.1016/j.bspc.2016.06.015
7. Aliev T, Quluyev Q, Pashayev F et al (2016) Intelligent seismic-acoustic system for identifying the location of the areas of an expected earthquake. J Geosci Environ Prot 4(4):147–162. https://doi.org/10.4236/gep.2016.44018
8. Aliev TA, Abbasov AM, Guluyev QA et al (2013) System of robust noise monitoring of anomalous seismic processes. Soil Dyn Earthq Eng 53:11–25. https://doi.org/10.1016/j.soildyn.2012.12.013
9. Aliev TA (2017) Intelligent seismic-acoustic system for identifying the area of the focus of an expected earthquake. In: Zouaghi T (ed) Earthquakes: tectonics, hazard and risk mitigation. Intech, London, pp 293–315. https://doi.org/10.5772/65403
10. Collacott RA (1977) Mechanical fault diagnosis and condition monitoring. Springer, Dordrecht. https://doi.org/10.1007/978-94-009-5723-7
11. Bendat JS, Piersol AG (2010) Random data: analysis and measurement procedures, 4th edn. Wiley, Hoboken. https://doi.org/10.1002/9781118032428.ch11
12. Proakis JG, Manolakis DG (2006) Digital signal processing: principles, algorithms, and applications, 4th edn. Pearson Prentice Hall, Upper Saddle River
13. Vetterli M, Kovacevic J, Goyal VK (2014) Foundations of signal processing, 3rd edn. Cambridge University Press, Cambridge
14. Owen M (2012) Practical signal processing. Cambridge University Press, Cambridge
15. Kay SM (2013) Fundamentals of statistical signal processing, volume III: practical algorithm development, 1st edn. Prentice Hall, Westford
16. Smith S (2002) Digital signal processing: a practical guide for engineers and scientists, 1st edn. Newnes, Amsterdam
17. Manolakis DG, Ingle VK (2011) Applied digital signal processing: theory and practice, 1st edn. Cambridge University Press, Cambridge. https://doi.org/10.1017/cbo9780511835261
18. Pashayev AM, Aliev TA, Hazarkhanov AT et al (2015) Method of preflight monitoring of the technical condition of aviation equipment. Eurasian Patent 021,468, 30 June 2015
19. Aliev TA, Musaeva NF, Guluyev GA et al (2011) System of monitoring of period of hidden transition of compressor station to emergency state. J Autom Inf Sci 43(11):66–81. https://doi.org/10.1615/JAutomatInfScien.v43.i11.70
20. Aliev TA, Musaeva NF, Nusratov OQ et al (2016) Models for indicating the period of failure of industrial objects. In: Zadeh L, Abbasov A, Abbasov A, Yager R et al (eds) Recent developments and new direction in soft-computing foundations and applications. Studies in fuzziness and soft computing, vol 342. Springer, Cham, pp 389–405. https://doi.org/10.1007/978-3-319-32229-2_27

References

21. Aliev TA, Abbasov AM, Aliev ER et al (2007) Digital technology and systems for generating and analyzing information from deep strata of the earth for the purpose of interference monitoring of the technical state of major structures. Autom Control Comput Sci 41(2):59–67. https://doi.org/10.3103/S0146411607020010
22. Aliev TA, Aliev ER (2008) Multichannel telemetric system for seismo-acoustic signal interference monitoring of earthquakes. Autom Contr Comput Sci 42(4):223–228. https://doi.org/10.3103/S0146411608040093
23. Aliev TA, Musayeva NF, Sattarova UE (2014) Noise technologies for operating the system for monitoring of the beginning of violation of seismic stability of construction objects. In: Zadeh L, Abbasov A, Yager R et al (eds) Recent developments and new directions in soft computing. Studies in fuzziness and soft computing, vol 317. Springer, Cham, pp 211–232. https://doi.org/10.1007/978-3-319-06323-2_14
24. Aliev TA, Musayeva NF, Suleymanova MT et al (2015) Analytical representation of the density function of normal distribution of noise. J Autom Inf Sci 47(8):24–40. https://doi.org/10.1615/JAutomatInfScien.v47.i8.30
25. Aliev TA, Musayeva NF, Sattarova UE et al (2013) The technology of forming the normalized correlation matrices of the matrix equations of multidimensional stochastic objects. J Autom Inf Sci 45(1):1–15. https://doi.org/10.1615/JAutomatInfScien.v45.i1.10
26. Aliev TA, Alizade AA, Etirmishli GD et al (2011) Intelligent seismoacoustic system for monitoring the beginning of anomalous seismic process. Seismic Instrum 47(1):15–23. https://doi.org/10.3103/S0747923911010026
27. Aliev TA, Abbasov AM, Mamedova GG et al (2013) Technologies for noise monitoring of abnormal seismic processes. Seismic Instru 49(1):64–80. https://doi.org/10.3103/S0747923913010015
28. Aliev TA (2000) Robust technology for systems analysis of seismic signals. Autom Contr Comput Sci 34(5):17–26
29. Aliev TA, Abbasov AM (2005) Digital technology and a system of interference monitoring of the technical state of physical structures and warnings of anomalous seismic processes. Autom Contr Comput Sci 39(6):1–7
30. Aliev TA (2001) A robust system for checking marine oil deposit seismic stability. Autom Contr Comput Sci 35(2):1–7
31. Aliev TA (2002) Algorithms and methodology for analysis of interference as an information-carrying medium. Autom Contr Comput Sci 36(1):1–10
32. Aliev TA (2004) Theoretical foundations of interference analysis of noisy signals. Autom Contr Comput Sci 38(3):18–27
33. Aliev TA (2008) Theoretical fundamentals of interference analysis and failure prediction. Cybern Syst Anal 44(4):482–492. https://doi.org/10.1007/s10559-008-9020-1
34. Aliev T, Aliev E (2007) Technology of noise analysis and monitoring of defect origin. Appl Comput Math 6(2):246–252
35. Aliev TA, Guluyev GA (2003) Information system for diagnostics and interference prediction of failures at compressor stations. Autom Contr Comput Sci 37(6):28–33
36. Aliev TA (2002) Theory of interference analysis. Autom Contr Comput Sci 36(6):29–39
37. Aliev TA (2003) Theory and technology of interference-enabled prediction of system breakdowns. Autom Contr Comput Sci 37(3):15–22
38. Aliev TA (2000) Algorithms for improving functional robustness of diagnostic systems. Autom Contr Comput Sci 34(1):43–50
39. Aliev TA, Rzaeva NE (2017) Techniques for determining robust estimates of correlation functions for random noisy signals. Meas Tech 60(4):343–349. https://doi.org/10.1007/s11018-017-1199-y
40. Aliev TA, Guliev GA, Pashayev FH et al (2011) Algorithms for determining the coefficient of correlation and cross-correlation function between a useful signal and noise of noisy technological parameters. Cybern Syst Anal 47(3):481–489. https://doi.org/10.1007/s10559-011-9329-z

41. Aliev TA (2001) Robust methodology for improving the conditionedness of correlation matrices. Autom Contr Comput Sci 35(1):8–20
42. Aliev TA, Guluyev GA, Rzayev AH et al (2009) Correlation indicators of microchanges in technical states of control objects. Cybern Syst Anal 45(4):655–662. https://doi.org/10.1007/s10559-009-9132-2
43. Aliev TA, Aliev ER, Mastalieva DI et al (2008) Adaptive technology of the sampling of noise-corrupted signals. Autom Contr Comput Sci 42(1):20–25
44. Aliev TA, Musaeva NF, Sattarova UE (2010) Robust technologies for calculating normalized correlation functions. Cybern Syst Anal 46(1):153–166. https://doi.org/10.1007/s10559-010-9194-1
45. Aliev TA, Musaeva NF, Sattarova UE (2011) Technology of calculating robust normalized correlation matrices. Cybern Syst Anal 47(1):152–165. https://doi.org/10.1007/s10559-011-9298-2
46. Aliev TA, Amirov ZA (1998) Algorithm to choose the regularization parameters for statistical identification. Autom Remote Contr 59(6):859–866
47. Aliev TA, Musaeva NF (1995) Algorithm for reducing errors in estimation of a correlation function of a noisy signal. Optoelectron Instrum Data Process 4:100–106
48. Aliev TA, Musaeva NF (1998) An algorithm for eliminating microerrors of noise in the solution of statistical dynamics problems. Autom Remote Contr 59(5):679–688
49. Aliev TA, Ali-Zade TA (1999) Robust algorithms for spectral analysis of the technological parameters of industrial plants. Autom Contr Comput Sci 33(5):38–44
50. Aliev TA, Guliev QA, Rzaev AH et al (2009) Position-binary and spectral indicators of microchanges in the technical states of control objects. Autom Contr Comput Sci 43(3):156–165. https://doi.org/10.3103/S0146411609030067
51. Aliev TA (2001) A robust technology for improving correlation and spectral characteristic estimators, correlation matrix conditioning, and identification adequacy. Autom Contr Comput Sci 35(4):10–19
52. Aliev TA, Alizadeh TA (2000) Robust technology for calculation of the coefficients of the Fourier series of random signals. Autom Contr Comput Sci 34(4):18–26
53. Aliev TA, Musaeva NF, Suleymanova MT et al (2016) Technology for calculating the parameters of the density function of the normal distribution of the useful component in a noisy process. J Autom Inf Sci 48(4):39–55. https://doi.org/10.1615/JAutomatInfScien.v48.i4.50
54. Aliev TA, Musaeva NF, Suleymanova MT (2017) Density function of noise distribution as an indicator for identifying the degree of fault growth in sucker rod pumping unit (SRPU). J Autom Inf Sci 49(4):1–11. https://doi.org/10.1615/JAutomatInfScien.v49.i4.10
55. Aliyev TA, Musayeva NF (1996) Statistical identification with error balancing. J Comput Syst Sci Int 34(5):119–124
56. Aliev TA, Musaeva NF (1997) Algorithms for determining a dispersion and errors caused by random signal interferences. Optoelectron Instrum Data Process 3:74–86
57. Aliev TA, Musaeva NF, Gazizade BI (2017) Algorithms for constructing a model of the noisy process by correcting the law of its distribution. J Autom Inf Sci 49(9):61–75. https://doi.org/10.1615/JAutomatInfScien.v49.i9.50
58. Aliev TA, Musaeva NF (1997) Algorithms for improving adequacy of statistical identification. J Comput Syst Sci Int 36(3):363–369
59. Aliev TA, Musaeva NF (1998) Algorithms providing adequacy to regression equations. Optoelectron Instrum Data Process 2:92–95
60. Aliev TA, Nusratov OK (1998) Algorithms for the analysis of cyclic signals. Autom Contr Comput Sci 32(2):59–64
61. Aliev TA, Nusratov OK (1998) Position-width-impulse analysis of cyclic and random signals. Optoelectron Instrum Data Proces 2:85–91
62. Aliev TA, Nusratov OK (1998) Position-width-pulse analysis and sampling of random signals. Autom Contr Comput Sci 32(5):44–48

63. Aliev TA, Nusratov OK (1998) Pulse-width and position method of diagnostics of cyclic processes. J Comput Syst Sci Int 37(1):126–131
64. Aliev TA, Mamedova UM (2003) Positional binary methodology for extraction of interference from noisy signals. Autom Contr Comput Sci 37(2):12–19

Chapter 2
Correlation Technology for Noise Control of the Beginning and Development Dynamics of the Latent Period of Accidents

Abstract The shortcomings of traditional technologies for correlation analysis of noisy signals at the beginning of the latent period of transition into an emergency condition are analyzed. It is shown that the noise correlated with the useful signal that appears during this period causes additional errors. At the same time, development dynamics of an accident in its latent period affects the duration of the correlation time and the degree of correlation between the useful signal and the noise. Noise technologies are proposed for calculating the estimates of the noise variance and the cross-correlation function between the useful signal and the noise, both for the case when the useful signal is correlated with the noise, and for the case when there is no correlation between them. Also proposed is the relay correlation technology for noise control of the development dynamics of accidents and the noise signaling of the beginning of the latent period of accidents at control objects. The possibility of improving the accuracy of the results of existing methods for analyzing noisy random signals with the use of these technologies is considered.

2.1 Correlation Noise Technology for Calculating the Variance of Noise of Noisy Signals in the Latent Period of Accidents

Many algorithms and technologies have been proposed for the purpose of eliminating the errors caused by various factors that affect the result of calculating the estimates of the statistical characteristics of noisy signals received from control objects over a protracted period. These algorithms and technologies are aimed at increasing the accuracy of calculating estimates of the statistical characteristics of noisy signals [1–12].

It was believed that there was no correlation between the noise and the useful signal. However, in the latent period of the emergency state of real objects, noises of technological parameters arise as a result of the beginning of these accidents. Because of this, a correlation emerges between the noise and the useful signal. Therefore, in these cases, on the one hand, the estimates obtained by traditional technologies contain a significant error. On the other hand, valuable diagnostic information contained in the noise is lost.

In the following paragraphs, we consider one of the possible technologies for calculating the estimate of the noise variance in the presence of a correlation between the useful signal and the noise of noisy random signals.

Let us consider this issue in more detail. It is known that when solving control problems in various industries, it becomes necessary to calculate the estimates of the auto- and cross-correlation functions of input (output) signals of control objects from expressions

$$R_{XX}(\mu) \approx \frac{1}{N} \sum_{k=1}^{N} X(i\Delta t) X((i+\mu)\Delta t), \qquad (2.1)$$

$$R_{XY}(\mu) \approx \frac{1}{N} \sum_{k=1}^{N} X(i\Delta t) Y((i+\mu)\Delta t), \qquad (2.2)$$

where $R_{XX}(\mu)$, $R_{XY}(\mu)$ are the estimates of the auto-correlation function of the input signal $X(i\Delta t)$ and the cross-correlation function between the input $X(i\Delta t)$ and the output $Y(i\Delta t)$ signals.

For real control objects, it is impossible to calculate the estimates of the correlation functions of the useful signals $X(i\Delta t)$ and $Y(i\Delta t)$ contaminated by the noises $\varepsilon(i\Delta t)$ and $\varphi(i\Delta t)$ from Expressions (2.1) and (2.2). This is due to the fact that the real input (output) signals of objects are sums of the centered useful signals $X(i\Delta t)$, $Y(i\Delta t)$ and the noises $\varepsilon(i\Delta t)$, $\eta(i\Delta t)$

$$\left. \begin{array}{l} g(i\Delta t) = X(i\Delta t) + \varepsilon(i\Delta t) \\ \eta(i\Delta t) = Y(i\Delta t) + \varphi(i\Delta t) \end{array} \right\} \qquad (2.3)$$

Taking into account (2.3), Algorithms (2.1) and (2.2) take on the following form:

$$R_{gg}(\mu) \approx \frac{1}{N} \sum_{i=1}^{N} g(i\Delta t) g((i+\mu)\Delta t)$$

$$= \frac{1}{N} \sum_{i=1}^{N} (X(i\Delta t) + \varepsilon(i\Delta t)) \times (X((i+\mu)\Delta t) + \varepsilon((i+\mu)\Delta t)) \qquad (2.4)$$

$$R_{g\eta}(\mu) = \frac{1}{N} \sum_{i=1}^{N} g(i\Delta t) \eta((i+\mu)\Delta t) = \frac{1}{N} \sum_{i=1}^{N} (Y(i\Delta t)$$
$$+ \varepsilon(i\Delta t))(Y((i+\mu)\Delta t) + \varphi((i+\mu)\Delta t)) \qquad (2.5)$$

Therefore, the control of object's operation is carried out on the basis of the estimates of the correlation functions found from (2.4) and (2.5). However, in the latent period of the emergency state, the following errors occur:

2.1 Correlation Noise Technology for Calculating the Variance of ...

$$\lambda_{gg}(\mu) \approx \frac{1}{N}\sum_{i=1}^{N}[X(i\Delta t)\varepsilon((i+\mu)\Delta t) + \varepsilon(i\Delta t)X((i+\mu)\Delta t) + \varepsilon(i\Delta t)\varepsilon((i+\mu)\Delta t)]$$

$$\lambda_{g\eta}(\mu) \approx \frac{1}{N}\sum_{i=1}^{N}[X(i\Delta t)\varphi((i+\mu)\Delta t) + \varepsilon(i\Delta t)Y((i+\mu)\Delta t) + \varepsilon(i\Delta t)\varphi((i+\mu)\Delta t)]$$

and, because of this, the following inequalities take place:

$$\left.\begin{array}{l} R_{XX}(\mu) \neq R_{gg}(\mu) \\ R_{XY}(\mu) \neq R_{g\eta}(\mu) \end{array}\right\}$$

For this reason, in reality, in many cases, it is impossible to ensure the adequacy of the results of solving problems of control in the latent period of accident. The conducted studies have shown that in order to eliminate this shortcoming, it is expedient to create algorithms and technologies for calculating the variance of noise and the cross-correlation function between the useful signal and the noise of noisy signals both in the presence and in the absence of correlation [1–12].

As shown in §1.1, the sum noise is composed of the noise $\varepsilon_1(i\Delta t)$ caused by external factors and the noise $\varepsilon_2(i\Delta t)$ caused by the initiation of various defects $\varepsilon_1(i\Delta t)$. Assume that $g(i\Delta t) = X(i\Delta t) + \varepsilon(i\Delta t)$ is a sampled stationary random signal with the normal distribution law, consisting of the useful signal $X(i\Delta t)$ and the noise $\varepsilon(i\Delta t)$ with a mathematical expectation equal to zero.

In this case, the formula for calculating the estimate of the variance of the noisy is

$$D_g \approx R_{gg}(0) = \frac{1}{N}\sum_{i=1}^{N}g^2(i\Delta t) = \frac{1}{N}\sum[X(i\Delta t)$$

$$+ \varepsilon(i\Delta t)][X(i\Delta t) + \varepsilon(i\Delta t)] \approx \frac{1}{N}\sum_{i=1}^{N}X^2(i\Delta t)$$

$$+ 2\frac{1}{N}\sum_{i=1}^{N}X(i\Delta t)\varepsilon(i\Delta t) + \frac{1}{N}\sum_{i=1}^{N}\varepsilon(i\Delta t)\varepsilon(i\Delta t) \approx R_{XX}(0)$$

$$+ 2R_{X\varepsilon}(0) + R_{\varepsilon\varepsilon}(0). \tag{2.6}$$

Therefore, when the sum noise $\varepsilon(i\Delta t) = \varepsilon_1(i\Delta t) + \varepsilon_2(i\Delta t)$ is correlated with the useful signal $X(i\Delta t)$, its variance D_ε is calculated from the following expression:

$$D_\varepsilon = 2R_{X\varepsilon}(0) + R_{\varepsilon\varepsilon}(0), \tag{2.7}$$

where $R_{X\varepsilon}(0) \approx \frac{1}{N}\sum_{i=1}^{N} X(i\Delta t)\varepsilon(i\Delta t)$ is the cross-correlation function between the useful signal and the noise; $R_{\varepsilon\varepsilon}(0) \approx \sum_{i=1}^{N} \varepsilon(i\Delta t)\varepsilon(i\Delta t)$ is the estimate of the variance of the noise $\varepsilon(i\Delta t)$.

The formula for calculating the estimate of $R_{gg}(\mu)$ can be represented similarly to Expression (2.6) in the form

$$R_{gg}(\mu) \approx \frac{1}{N}\sum_{i=1}^{N} g(i\Delta t)g((i+\mu)\Delta t) \approx \frac{1}{N}\sum_{i=1}^{N}[X(i\Delta t)$$

$$+ \varepsilon(i\Delta t)] \times [X((i+\mu)\Delta t) + \varepsilon((i+\mu)\Delta t)] \approx \frac{1}{N}\sum_{i=1}^{N}[X(i\Delta t)X((i+\mu)\Delta t)$$

$$+ \varepsilon(i\Delta t)X((i+\mu)\Delta t) + X(i\Delta t)\varepsilon((i+\mu)\Delta t)$$
$$+ \varepsilon(i\Delta t)\varepsilon((i+\mu)\Delta t)] \approx R_{XX}(\mu) + R_{\varepsilon X}(\mu) + R_{X\varepsilon}(\mu) + R_{\varepsilon\varepsilon}(\mu)$$

$$\approx \begin{cases} R_{XX}(0) + 2R_{X\varepsilon}(0) + R_{\varepsilon\varepsilon}(0) & \text{when } \mu = 0 \\ R_{XX}(\mu) + 2R_{X\varepsilon}(\mu) & \text{when } \mu \neq 0 \end{cases} \quad (2.8)$$

It follows from (2.6)–(2.8) that, without taking into account the noise characteristics, the estimates of the correlation functions $R_{X\varepsilon}(0)$ and $R_{\varepsilon\varepsilon}(0)$ will have errors $\lambda_{gg}(\mu)$. Experimental studies have shown that for real control objects in the latent period of an emergency state, often, as a result of accident development dynamics, even during several sampling intervals, i.e., at $\mu = \Delta t, 2\Delta t, 3\Delta t \ldots$, there is a correlation between $X(i\Delta t)$ and $\varepsilon(i\Delta t)$. Therefore, in order to eliminate the error $\lambda_{gg}(\mu)$ in the estimate of $R_{gg}(\mu)$, in addition to calculating the estimate of D_ε for $\mu = 0$, it is also required to develop a technology for calculating the estimates of the cross-correlation functions between $X(i\Delta t)$ and $\varepsilon(i\Delta t)$: $R_{X\varepsilon}(\Delta t), R_{X\varepsilon}(2\Delta t), R_{X\varepsilon}(3\Delta t), \ldots$ when $\Delta t = \Delta t_\varepsilon$ [1].

It is known [2] that when the sampling interval decreases, the estimates of $R_{XX}(\mu)$ at $\mu = 0, \Delta t, 2\Delta t$ become close values and their difference turns out to be commensurable with the ΔX level quantization interval, which is determined by the measurement resolution. For instance, for an analog-to-digital converter, it is equal to the weight of the least significant bit [13]. As the observation time T tends to infinity and the sampling interval $\Delta t \to \Delta t_\varepsilon$, the estimates of $R_{XX}(\mu = 0)$ and $R_{XX}(\mu = \Delta t)$ turn out to be so close that we can consider the following inequalities true:

$$\lim_{\substack{T\to\infty \\ \Delta t \to 0}}[R_{XX}(\mu = 0) - R_{XX}(\mu = \Delta t)] \ll \Delta X \quad (2.9)$$

$$\lim_{\substack{T\to\infty \\ \Delta t \to 0}}[R_{XX}(\mu = \Delta t) - R_{XX}(\mu = 2\Delta t)] \ll \Delta X \quad (2.10)$$

In this case, assuming that the equalities

$$\left. \begin{aligned} \frac{1}{N}\sum_{i=1}^{N} X(i\Delta t)\varepsilon((i+1)\Delta t) \approx 0 \\ \frac{1}{N}\sum_{i=1}^{N} X(i\Delta t)\varepsilon((i+2)\Delta t) \approx 0 \end{aligned} \right\}$$

2.1 Correlation Noise Technology for Calculating the Variance of ...

$$\left.\begin{array}{l} R_{gg}(\mu = 0) \approx R_{xx}(\mu = 0) + D_\varepsilon \\ R_{gg}(\mu = 1) \approx R_{xx}(\mu = 1) \\ R_{gg}(\mu = 2) \approx R_{xx}(\mu = 2) \end{array}\right\}$$

true and taking into account the conditions

$$[R_{XX}(0) + D_\varepsilon] - R_{XX}(0) \neq R_{XX}(\mu = \Delta t) - R_{XX}(\mu = 2\Delta t),$$
$$R_{XX}(\mu = 0) - R_{XX}(\mu = \Delta t) \neq R_{XX}(\mu = \Delta t) - R_{XX}(\mu = 2\Delta t),$$
$$R_{gg}(\mu = 0) - R_{gg}(\mu = \Delta t) \neq R_{gg}(\mu = \Delta t) - R_{gg}(\mu = 2\Delta t).$$

The following can be written:

$$D_\varepsilon \approx R_{gg}(\mu = 0) + R_{gg}(\mu = 2) 2 R_{gg}(\mu = 1)$$
$$= \frac{1}{N}\sum_{i=1}^{N} g(i\Delta t)g(i\Delta t) - \frac{1}{N}\sum_{i=1}^{N} 2g(i\Delta t)g((i+1)\Delta t)$$
$$+ \frac{1}{N}\sum_{i=1}^{N} g(i\Delta t)g((i+2)\Delta t) = \frac{1}{N}\sum_{i=1}^{N} [g(i\Delta t)g(i\Delta t)$$
$$- 2g(i\Delta t)g((i+1)\Delta t) + g(i\Delta t)g((i+2)\Delta t)] \quad (2.11)$$

Thus, when these conditions are fulfilled, Expression (2.11) for calculating the variance D_ε of the noise $\varepsilon(i\Delta t)$ can be represented as

$$D_\varepsilon \approx \frac{1}{N}\sum_{i=1}^{N} \left[g^2(i\Delta t) - 2g(i\Delta t)g((i+1)\Delta t) + g(i\Delta t)g((i+2)\Delta t) \right] \quad (2.12)$$

The validity of this expression can be verified by decomposing its right-hand side into the corresponding summands, i.e.,

$$D_\varepsilon \approx \frac{1}{N}\sum_{i=1}^{N} [g(i\Delta t)g(i\Delta t) - 2g(i\Delta t)g((i+1)\Delta t) + g(i\Delta t)g((i+2)\Delta t)]$$
$$\approx \frac{1}{N}\sum_{i=1}^{N} [X(i\Delta t) + \varepsilon(i\Delta t)][X(i\Delta t) + \varepsilon(i\Delta t)] - \frac{1}{N}\sum_{i=1}^{N} 2[X(i\Delta t)$$
$$+ \varepsilon(i\Delta t)][X((i+1)\Delta t) + \varepsilon((i+1)\Delta t)]$$
$$+ \frac{1}{N}\sum_{i=1}^{N} [X(i\Delta t) + \varepsilon(i\Delta t)][X((i+2)\Delta t) + \varepsilon((i+2)\Delta t)]$$
$$= \frac{1}{N}\sum_{i=1}^{N} [g(i\Delta t)g(i\Delta t) - 2g(i\Delta t)g((i+1)\Delta t) + g(i\Delta t)g((i+2)\Delta t)]$$

$$\approx \frac{1}{N} \sum_{i=1}^{N} [X(i\Delta t) + \varepsilon(i\Delta t)][X(i\Delta t) + \varepsilon(i\Delta t)]$$

$$- \frac{1}{N} \sum_{i=1}^{N} 2[X(i\Delta t) + \varepsilon(i\Delta t)] \times [X((i+1)\Delta t) + \varepsilon((i+1)\Delta t)]$$

$$+ \frac{1}{N} \sum_{i=1}^{N} [X(i\Delta t) + \varepsilon(i\Delta t)][X((i+2)\Delta t) + \varepsilon((i+2)\Delta t)]$$

$$= R_{XX}(0) + R_{X\varepsilon}(0) + R_{\varepsilon X}(0) + R_{\varepsilon\varepsilon}(0) - 2R_{cXX}(\Delta t) - 2R_{X\varepsilon}(\Delta t)$$
$$- 2R_{\varepsilon X}(\Delta t) - 2R_{\varepsilon\varepsilon}(\Delta t) + R_{XX}(2\Delta t) + R_{X\varepsilon}(2\Delta t) + R_{\varepsilon X}(2\Delta t) + R_{\varepsilon\varepsilon}(2\Delta t).$$

In this case, if the stationarity conditions and the normalcy of the distribution of noisy signals are fulfilled, then the following equalities will be true:

$$R_{X\varepsilon}(0) \approx \frac{1}{N} \sum_{i=1}^{N} X(i\Delta t)\varepsilon(i\Delta t) \neq 0,$$

$$R_{\varepsilon X} \approx \frac{1}{N} \sum_{i=1}^{N} \varepsilon(i\Delta t)X(i\Delta t) \neq 0,$$

$$R_{\varepsilon\varepsilon} \approx \frac{1}{N} \sum_{i=1}^{N} \varepsilon(i\Delta t)\varepsilon(i\Delta t) \neq 0,$$

$$R_{XX}(0) + R_{XX}(2\Delta t) - 2R_{XX}(\Delta t) \approx 0,$$

$$R_{\varepsilon\varepsilon}(\Delta t) \approx \frac{1}{N} \sum_{i=1}^{N} \varepsilon(i\Delta t)\varepsilon((i+1)\Delta t) \approx 0,$$

$$R_{\varepsilon\varepsilon}(2\Delta t) \approx \frac{1}{N} \sum_{i=1}^{N} \varepsilon(i\Delta t)\varepsilon((i+2)\Delta t) \approx 0,$$

$$R_{X\varepsilon}(\Delta t) \approx \frac{1}{N} \sum_{i=1}^{N} X(i\Delta t)\varepsilon((i+1)\Delta t) \approx 0,$$

$$R_{X\varepsilon}(2\Delta t) \approx \frac{1}{N} \sum_{i=1}^{N} X(i\Delta t)\varepsilon((i+2)\Delta t) \approx 0,$$

$$R_{\varepsilon X}(\Delta t) \approx \frac{1}{N} \sum_{i=1}^{N} \varepsilon(i\Delta t)X((i+1)\Delta t) \approx 0,$$

$$R_{\varepsilon X}(2\Delta t) \approx \frac{1}{N} \sum_{i=1}^{N} \varepsilon(i\Delta t)X((i+2)\Delta t) \approx 0.$$

Consequently, in the right-hand side of Eq. (2.12), we obtain

$$D_\varepsilon \approx R_{X\varepsilon}(0) + R_{\varepsilon X}(0) + R_{\varepsilon\varepsilon}(0) \approx 2R_{X\varepsilon}(0) + R_{\varepsilon\varepsilon}(0) \quad (2.13)$$

Thus, the estimate obtained from (2.9) and (2.10) is the estimate of the noise variance. The formula for calculating it can be represented as

$$D_\varepsilon \approx \frac{1}{N} \sum_{i=1}^{N} \left[g^2(i\Delta t) + g(i\Delta t)g((i+2)\Delta t) - 2g(i\Delta t)g((i+1)\Delta t)\right] \quad (2.14)$$

2.2 Correlation Technology of Noise Control of Accident Development Dynamics

As shown above, at the beginning of the latent period of accident initiation, due to the emergence of the noise $\varepsilon_2(i\Delta t)$, the estimate of the variance of the total noise $\varepsilon(i\Delta t)$ differs from zero. In this case, with a stable initial emergency condition, this estimate does not change. However, with as the defect develops, this estimate increases. Because of this, it becomes possible to control accident development dynamics. However, as is obvious from Expressions (2.12) and (2.13), the estimate of the noise variance is formed by the sum of the variance of the noise $\varepsilon(i\Delta t)$ caused by external factors and the noise $\varepsilon_2(i\Delta t)$ caused by defect initiation. Therefore, it is not possible to control accident development dynamics on the basis of an increase or a decrease in the value of the estimates of the noise. As numerous experiments demonstrate, accident development dynamics leads both to an increase in the variance of the noise $\varepsilon_2(i\Delta t)$ and to an increase in the duration of the time correlation. As a result of the development dynamics, a correlation emerges between at the beginning $X(i\Delta t)$ and $\varepsilon(i+1)\Delta t$. Further development of the dynamics leads to the emergence of a correlation between $X(i\Delta t)$ and $\varepsilon(i+2)\Delta t$, then between $X(i\Delta t)$ and $\varepsilon(i+3)\Delta t$, etc. Therefore, in order to control accident development dynamics, it is necessary to calculate the estimates that correspond to the cross-correlation function between $X(i\Delta t)$ and $\varepsilon(i\Delta t)$. In this regard, let us consider the possibility of calculating the estimate of $R_{X\varepsilon}(\Delta t)$ in the presence of a correlation between $X(i\Delta t)$ and $\varepsilon(i\Delta t)$ for $\mu = \Delta t$ from the expression

$$R'_{X\varepsilon}(\mu = \Delta t) \approx \frac{1}{N} \sum_{i=1}^{N} [g(i\Delta t)g(i+1) - 2g(i\Delta t)g((i+2)\Delta t) + g(i\Delta t)g((i+3)\Delta t)]$$

$$\approx \frac{1}{N} \sum_{i=1}^{N} [g(i\Delta t)g((i+1)\Delta t)]$$

$$- \frac{1}{N} \sum_{i=1}^{N} 2[g(i\Delta t)g((i+2)\Delta t)] + \frac{1}{N} \sum_{i=1}^{N} [g(i\Delta t)g((i+3)\Delta t)]$$

$$\approx \frac{1}{N} \sum_{i=1}^{N} [X(i\Delta t) + \varepsilon(i\Delta t)][X((i+1)\Delta t) + \varepsilon((i+1)\Delta t)]$$

$$-\frac{1}{N}\sum_{i=1}^{N}2[X(i\Delta t)+\varepsilon(i\Delta t)]\times[X((i+2)\Delta t)+\varepsilon((i+2)\Delta t)]$$

$$+\frac{1}{N}\sum_{i=1}^{N}[X(i\Delta t)+\varepsilon(i\Delta t)][X((i+3)\Delta t)+\varepsilon((i+3)\Delta t)]$$

$$\approx R_{XX}(\Delta t)+R_{X\varepsilon}(\Delta t)+R_{\varepsilon X}(\Delta t)+R_{\varepsilon\varepsilon}(\Delta t)-2R_{XX}(2\Delta t)$$
$$-2R_{X\varepsilon}(2\Delta t)-2R_{\varepsilon X}(2\Delta t)-2R_{\varepsilon\varepsilon}(2\Delta t)+R_{XX}(3\Delta t)+R_{X\varepsilon}(3\Delta t)$$
$$+R_{\varepsilon X}(3\Delta t)+R_{\varepsilon\varepsilon}(3\Delta t)$$

When the stationarity conditions and the normality of the distribution of noisy signals are fulfilled, assuming that the relations

$$R_{X\varepsilon}(\Delta t)\approx\frac{1}{N}\sum_{i=1}^{N}X(i\Delta t)\varepsilon((i+1)\Delta t)\neq 0,$$

$$R_{\varepsilon X}(\Delta t)\approx\frac{1}{N}\sum_{i=1}^{N}\varepsilon(i\Delta t)X((i+1)\Delta t)\neq 0,$$

$$R_{XX}(\Delta t)+R_{XX}(3\Delta t)-2R_{XX}(2\Delta t)\approx 0,$$

$$R_{\varepsilon\varepsilon}(\Delta t)\approx 0,\ R_{\varepsilon\varepsilon}(3\Delta t)\approx 0,\ R_{\varepsilon\varepsilon}(2\Delta t)\approx 0,$$

$$R_{\varepsilon\varepsilon}(\Delta t)+R_{\varepsilon\varepsilon}(3\Delta t)-2R_{\varepsilon\varepsilon}(2\Delta t)\approx 0,$$

$$R_{X\varepsilon}(2\Delta t)\approx\frac{1}{N}\sum_{i=1}^{N}X(i\Delta t)\varepsilon((i+2)\Delta t)\approx 0,$$

$$R_{X\varepsilon}(3\Delta t)\approx\frac{1}{N}\sum_{i=1}^{N}X(i\Delta t)\varepsilon((i+3)\Delta t)\approx 0,$$

$$R_{\varepsilon X}(2\Delta t)\approx\frac{1}{N}\sum_{i=1}^{N}\varepsilon(i\Delta t)X((i+2)\Delta t)\approx 0,$$

$$R_{\varepsilon X}(3\Delta t)\approx\frac{1}{N}\sum_{i=1}^{N}\varepsilon(i\Delta t)X((i+3)\Delta t)\approx 0$$

is true, we get the equality

$$R'_{X\varepsilon}(\Delta t)\approx R_{X\varepsilon}(\Delta t)+R_{\varepsilon X}(\Delta t)\approx 2R_{X\varepsilon}(\Delta t),$$

from which we can calculate the estimates of $R_{X\varepsilon}(\Delta t)$

$$R_{X\varepsilon}(\Delta t)\approx\frac{R'_{X\varepsilon}(\Delta t)}{2}\approx\frac{1}{N}\sum_{i=1}^{N}[g(i\Delta t)g(i\Delta t)-2g(i\Delta t)g((i+1)\Delta t)+g(i\Delta t)g((i+2)\Delta t)]$$

$$\approx\frac{1}{2N}\sum_{i=1}^{N}[g(i\Delta t)g((i+1)\Delta t)]-\frac{1}{N}\sum_{i=1}^{N}2[g(i\Delta t)g((i+2)\Delta t)]$$

2.2 Correlation Technology of Noise Control of Accident Development Dynamics

$$+ \frac{1}{N} \sum_{i=1}^{N} [g(i\Delta t)g((i+3)\Delta t)]$$

Obviously, the expression for calculating the estimate of $R_{X\varepsilon}(2\Delta t)$ in the presence of a correlation between $X(i\Delta t)$ and $\varepsilon(i\Delta t)$ for $\mu = 2\Delta t$ can also be written in a similar way. Therefore, in the presence of a correlation between $X(i\Delta t)$ and $\varepsilon(i\Delta t)$ for m different time shifts $\mu = m\Delta t$, $m = 1, 2, 3, \ldots$, the following generalized expression is true:

$$R'_{X\varepsilon}(m\Delta t) \approx \frac{1}{N} \sum_{i=1}^{N} [g(i\Delta t)g((i+m-1)\Delta t) - 2g(i\Delta t)g((i+m)\Delta t)$$

$$+ g(i\Delta t)g((i+m+1)\Delta t)] \approx \frac{1}{N} \sum_{i=1}^{N} [g(i\Delta t)g((i+m)\Delta t)]$$

$$- \frac{1}{N} \sum_{i=1}^{N} 2[g(i\Delta t)g((i+m+1)\Delta t)]$$

$$+ \frac{1}{N} \sum_{i=1}^{N} [g(i\Delta t)g((i+m+2)\Delta t)]$$

$$\approx \frac{1}{N} \sum_{i=1}^{N} [X(i\Delta t)\varepsilon(i\Delta t)][X((i+m)\Delta t) + \varepsilon((i+m)\Delta t)]$$

$$- \frac{1}{N} \sum_{i=1}^{N} 2[X(i\Delta t) + \varepsilon(i\Delta t)]$$

$$\times [X((i+m+1)\Delta t) + \varepsilon((i+m+1)\Delta t)] + \frac{1}{N} \sum_{i=1}^{N} [X(i\Delta t) + \varepsilon(i\Delta t)]$$

$$\times [X((i+m+2)\Delta t) + \varepsilon((i+m+2)\Delta t)] \approx R_{XX}(m\Delta t) + R_{X\varepsilon}(m\Delta t)$$
$$+ R_{\varepsilon X}(m\Delta t) + R_{\varepsilon\varepsilon}(m\Delta t) - 2R_{XX}((m+1)\Delta t)$$
$$- 2R_{X\varepsilon}((m+1)\Delta t) - 2R_{\varepsilon X}((m+1)\Delta t) - 2R_{\varepsilon\varepsilon}((m+1)\Delta t)$$
$$+ R_{XX}((m+2)\Delta t) + R_{X\varepsilon}((m+2)\Delta t) + R_{\varepsilon X}((m+2)\Delta t)$$
$$+ R_{\varepsilon\varepsilon}((m+2)\Delta t)$$

despite the fact that there is a correlation between $X(i\Delta t)$ and $\varepsilon(i\Delta t)$ at $\mu = m\Delta t$, $m = 1, 2, 3, \ldots$ and the following conditions are fulfilled

$$R_{X\varepsilon}(m\Delta t) \approx \frac{1}{N} \sum_{i=1}^{N} X(i\Delta t)\varepsilon((i+m)\Delta t) \neq 0,$$

$$R_{\varepsilon X}(m\Delta t) \approx \frac{1}{N} \sum_{i=1}^{N} \varepsilon(i\Delta t)X((i+m)\Delta t) \neq 0,$$

$$R_{XX}(m\Delta t) + R_{XX}((m+2)\Delta t) - 2R_{XX}((m+1)\Delta t) \approx 0,$$
$$R_{\varepsilon\varepsilon}(m\Delta t) \approx 0, \ R_{\varepsilon\varepsilon}((m+1)\Delta t) \approx 0, \ R_{\varepsilon\varepsilon}((m+2)\Delta t) \approx 0,$$
$$R_{\varepsilon\varepsilon}(m\Delta t) + R_{\varepsilon\varepsilon}((m+2)\Delta t) - 2R_{\varepsilon\varepsilon}((m+1)\Delta t) \approx 0,$$
$$R_{X\varepsilon}((m+1)\Delta t) \approx \frac{1}{N}\sum_{i=1}^{N} X(i\Delta t)\varepsilon((i+(m+1))\Delta t) \approx 0,$$
$$R_{X\varepsilon}((m+2)\Delta t) \approx \frac{1}{N}\sum_{i=1}^{N} X(i\Delta t)\varepsilon((i+(m+2))\Delta t) \approx 0,$$
$$R_{\varepsilon X}((m+1)\Delta t) \approx \frac{1}{N}\sum_{i=1}^{N} \varepsilon(i\Delta t)X((i+(m+1))\Delta t) \approx 0,$$
$$R_{\varepsilon X}((m+2)\Delta t) \approx \frac{1}{N}\sum_{i=1}^{N} \varepsilon(i\Delta t)X((i+(m+2))\Delta t) \approx 0.$$

Through this, we obtain the equality $R'_{X\varepsilon}(m\Delta t) \approx 2R_{X\varepsilon}(m\Delta t)$ that allows us to write the generalized expression for calculating $R_{X\varepsilon}(m\Delta t)$ as

$$R_{X\varepsilon}(m\Delta t) \approx \frac{1}{2}R'_{X\varepsilon}(m\Delta t) \approx \frac{1}{2N}\sum_{i=1}^{N}[g(i\Delta t)g((i+(m+1))\Delta t)$$
$$-2g(i\Delta t)g((i+(m+1))\Delta t) + g(i\Delta t)g((i+(m+2))\Delta t)],$$
$$R^R_{gg}(m\Delta t) = R_{gg}(m\Delta t) - R_{X\varepsilon}(m\Delta t).$$

Our experimental analysis of noisy signals received at compressor and seismic-acoustic stations, as well as on fixed offshore platforms, oil and gas production and refining facilities [1–12], showed that depending on the degree of accident development dynamics at these objects, a correlation appears between the useful signal $X(i\Delta t)$ and the noise $\varepsilon(i\Delta t)$ first at $\mu = 1\Delta t$, then at $\mu = 2\Delta t$, $\mu = 3\Delta t$, then at $\mu = 4\Delta t, 5\Delta t, 6\Delta t$, etc., and the values of these estimates reflect the accident development dynamics. Therefore, the generalized expression for calculating the estimates of $R_{X\varepsilon}(\mu = 1\Delta t), R_{X\varepsilon}(\mu = 2\Delta t), R_{X\varepsilon}(\mu = 3\Delta t), \ldots, R_{X\varepsilon}(\mu = m\Delta t)$ makes it possible to control not only the beginning but also the development dynamics of accidents.

2.3 Correlation Technology of Noise Signaling for the Beginning and Development Dynamics of the Latent Period of Accidents

As shown above, under normal operation mode, the noise $\varepsilon(i\Delta t) = \varepsilon_1(i\Delta t)$ arises during the period T_0 because of random external factors that do not corre-

2.3 Correlation Technology of Noise Signaling for the Beginning ...

late with the useful signal $X(i\Delta t)$. However, during the period T_1, t the beginning of the latent period of accidents, the noise $\varepsilon_2(i\Delta t)$ emerges, which is caused by the initiation of various defects. Therefore, at the starting moment of the period T_1, a correlation appears between the useful signal $X(i\Delta t)$ and the sum noise $\varepsilon(i\Delta t) = \varepsilon_1(i\Delta t) + \varepsilon_2(i\Delta t)$, and the estimate of the noise variance differs from zero. However, the obtained information does not allow us to determine the development dynamics of accidents. This is due to the fact that, in essence, the dynamics of the development of accidents, despite the indirect effect on the value of the estimate of the noise variance, manifests itself clearly only in the estimates of the cross-correlation functions between $X(i\Delta t)$ and $\varepsilon(i\Delta t)$ at various time shifts. Therefore, as mentioned above, to monitor the dynamics of the development of accidents, it is advisable to use the estimate $R_{X\varepsilon}(m)$. However, signaling the onset and development of a time correlation between the useful signal $X(i\Delta t)$ and the noise $\varepsilon(i\Delta t)$ is quite often sufficient to control the beginning and development dynamics of the latent period of accidents. Therefore, in addition to complex accident dynamics control technologies, it is also advisable to use in control systems simple technologies for signaling the start of this process. From this point of view, as an informative attribute for controlling accident development dynamics, we should use the estimates of the relay correlation functions, which can be calculated using the following formula:

$$R_{X\varepsilon}^*(\mu) = \frac{1}{N} \sum_{i=1}^{N} \operatorname{sgn} g(i\Delta t)\varepsilon(i\Delta t).$$

However, to use this formula, it is necessary to determine the samples of the noise $\varepsilon(i\Delta t)$ that cannot be measured directly.

In view of the above, let us consider one of the possible variants of approximate calculation of the estimates of the relay cross-correlation function $R_{X\varepsilon}^*(\mu)$ between the useful signal $X(i\Delta t)$ and the noise $\varepsilon(i\Delta t)$.

To this end, we first take the following notation and conditions:

$$\operatorname{sgn} g(i\Delta t) = \begin{cases} +1 & \text{when } g(i\Delta t) > 0 \\ 0 & \text{when } g(i\Delta t) = 0 \\ -1 & \text{when } g(i\Delta t) < 0 \end{cases}$$

In this case, the expression for calculating $R_{gg}^*(\mu)$ can be represented in the following form:

$$R_{gg}^*(\mu = 0) = \frac{1}{N} \sum_{i=1}^{N} \operatorname{sgn} g(i\Delta t) \cdot g(i\Delta t)$$

$$= \frac{1}{N} \sum_{i=1}^{N} \operatorname{sgn} X(i\Delta t) \cdot [X(i\Delta t) + \varepsilon(i\Delta t)]$$

$$= \frac{1}{N} \sum_{i=1}^{N} \operatorname{sgn} X(i\Delta t) \cdot X(i\Delta t) + \frac{1}{N} \sum_{i=1}^{N} \operatorname{sgn} X(i\Delta t) \cdot \varepsilon(i\Delta t)$$

$$= \operatorname{sgn} R_{xx}(0) + \operatorname{sgn} R_{x\varepsilon}(0).$$

The literature [1, 2] shows that the estimates of $R_{X\varepsilon}(0)$ can be calculated from the expression

$$R^*_{X\varepsilon}(\mu = 0) = R^*_{gg}(\mu = 0) + R^*_{gg}(\mu = 2) - 2R^*_{gg}(\mu = 1)$$

$$= \frac{1}{N} \sum_{i=1}^{N} \operatorname{sgn} g(i\Delta t) g(i\Delta t) - \frac{1}{N} \sum_{i=1}^{N} 2\operatorname{sgn} g(i\Delta t) g((i+1)\Delta t)$$

$$+ \frac{1}{N} \sum_{i=1}^{N} \operatorname{sgn} g(i\Delta t) g((i+2)\Delta t) = \frac{1}{N} \sum_{i=1}^{N} [\operatorname{sgn} g(i\Delta t) g(i\Delta t)$$

$$- 2\operatorname{sgn} g(i\Delta t) g((i+1)\Delta t) + \operatorname{sgn} g(i\Delta t) g((i+2)\Delta t)] \qquad (2.15)$$

Taking into account the equalities

$$\begin{cases} \operatorname{sgn} g(i\Delta t) = \operatorname{sgn} X(i\Delta t) \\ \operatorname{sgn} g(i\Delta t) \cdot g(i\Delta t) = \operatorname{sgn} X(i\Delta t) \cdot [X(i\Delta t) + \varepsilon(i\Delta t)] \\ R^*_{gg}(0) = R^*_{xx}(0) + R^*_{x\varepsilon}(0) \end{cases}$$

we can verify the validity of this formula by expanding its right-hand side into terms, i.e.,

$$R^*_{x\varepsilon}(\mu = 0) \approx \frac{1}{N} \sum_{i=1}^{N} \big[\operatorname{sgn} g(i\Delta t) g(i\Delta t) - 2 \operatorname{sgn} g(i\Delta t)$$

$$\times g((i+1)\Delta t) + \operatorname{sgn} g(i\Delta t) g((i+2)\Delta t) \big]$$

$$\approx \frac{1}{N} \sum_{i=1}^{N} \operatorname{sgn}[X(i\Delta t)][X(i\Delta t) + \varepsilon(i\Delta t)]$$

$$- \frac{1}{N} \sum_{i=1}^{N} 2 \operatorname{sgn}[X(i\Delta t)]$$

$$\times [X((i+1)\Delta t) + \varepsilon((i+1)\Delta t)]$$

$$+ \frac{1}{N} \sum_{i=1}^{N} \operatorname{sgn}[X(i\Delta t)][X((i+2)\Delta t) + \varepsilon((i+2)\Delta t)]$$

$$= \frac{1}{N} \sum_{i=1}^{N} \operatorname{sgn} X(i\Delta t) X(i\Delta t) + \frac{1}{N} \sum_{i=1}^{N} \operatorname{sgn} X(i\Delta t) \varepsilon(i\Delta t)$$

$$+ \frac{1}{N}\sum_{i=1}^{N} 2\operatorname{sgn} X(i\Delta t)X(i+1)\Delta t$$

$$-\frac{1}{N}\sum_{i=1}^{N} 2\operatorname{sgn} X(i+1)\Delta t + \frac{1}{N}\sum_{i=1}^{N}\operatorname{sgn} X(i\Delta t)X(i+2)\Delta t$$

$$+\frac{1}{N}\sum_{i=1}^{N}\operatorname{sgn} X(i\Delta t)\varepsilon(i+2)\Delta t = R_{XX}^{*}(0) + R_{X\varepsilon}^{*}(0)$$

$$- 2R_{XX}^{*}(\Delta t) - 2R_{X\varepsilon}^{*}(\Delta t) + R_{XX}^{*}(2\Delta t) + R_{X\varepsilon}^{*}(2\Delta t) \quad (2.16)$$

If the conditions of stationarity and the normality of the distribution of noisy signals are fulfilled, then the following equalities will be true for control objects:

$$\left.\begin{array}{l} R_{X\varepsilon}^{*}(0) = \frac{1}{N}\sum_{i=1}^{N}\operatorname{sgn} X(i\Delta t)\varepsilon(i\Delta t) \neq 0 \\ R_{XX}^{*}(0) + R_{XX}^{*}(2\Delta t) - 2R_{XX}^{*}(\Delta t) \approx 0 \end{array}\right\}$$

$$\left.\begin{array}{l} R_{X\varepsilon}^{*}(\Delta t) = \frac{1}{N}\sum_{i=1}^{N}\operatorname{sgn} X(i\Delta t)\varepsilon((i+1)\Delta t) \approx 0 \\ R_{X\varepsilon}^{*}(2\Delta t) = \frac{1}{N}\sum_{i=1}^{N} X(i\Delta t)\varepsilon((i+2)\Delta t) \approx 0 \end{array}\right\}$$

and therefore the right-hand side of Expression (2.16) takes the form

$$R_{X\varepsilon}^{*}(\mu = 0) \approx R_{X\varepsilon}^{*}(0),$$

which confirms that the estimate obtained from expression (2.15) is the estimate of $R_{X\varepsilon}^{*}(0)$.

Taking into account the generalization of the expressions for calculating $R_{X\varepsilon}(\mu)$ for various time shifts $m\Delta t$ between $X(i\Delta t)$ and $\varepsilon(i\Delta t)$, the formula for calculating $R_{X\varepsilon}^{*}(\mu)$ for various $m\Delta t$ time shifts in the generalized form can be represented in the following form:

$$R_{X\varepsilon}^{*}(m\Delta t) \approx \frac{1}{2}R_{X\varepsilon}^{'}(m\Delta t) \approx \frac{1}{2N}\sum_{i=1}^{N}\bigl[\operatorname{sgn} g(i\Delta t)g((i+m)\Delta t)$$

$$- 2\operatorname{sgn} g(i\Delta t)g((i+(m+1))\Delta t)$$

$$+ \operatorname{sgn} g(i\Delta t)g((i+(m+2))\Delta t)\bigr]. \quad (2.17)$$

Thus, we can assume that the estimate obtained from formula (2.17) is the estimate of the relay correlation function of $R_{X\varepsilon}^{*}(\mu\Delta t)$ between the useful signal $X(i\Delta t)$ and the noise $\varepsilon(i\Delta t)$.

For instance, the formula for calculating the estimate of $R_{X\varepsilon}^{*}(\mu = 1\Delta t)$ at $\mu = 1\Delta t$ will be

$$R^*_{X\varepsilon}(\mu = 1\Delta t) \approx \frac{1}{N}\sum_{i=1}^{N}\Big[\text{sgn}\, g(i\Delta t)g(i+1)\Delta t$$
$$- 2\,\text{sgn}\, g(i\Delta t)g(i+2)\Delta t + \text{sgn}\, g(i\Delta t)g(i+3)\Delta t\Big].$$

The formula for calculating the estimate of $R^*_{X\varepsilon}(\mu = 2\Delta t)$ at $\mu = 2\Delta t$ will be

$$R^*_{X\varepsilon}(\mu = 2\Delta) \approx \frac{1}{N}\sum_{i=1}^{N}\Big[\text{sgn}\, g(i\Delta t)g(i+2)\Delta t$$
$$- \text{sgn}\, g(i\Delta t)g(i+3)\Delta t + \text{sgn}\, g(i\Delta t)g(i+4)\Delta t\Big].$$

It is obvious that the estimates of $R^*_{X\varepsilon}(\mu = 3\Delta t)$, $R^*_{X\varepsilon}(\mu = 4\Delta t)$, ..., can be calculated in a similar manner.

Consequently, the generalized formula for calculating the estimate of $R^*_{X\varepsilon}$ can be written as

$$R^*_{X\varepsilon}(m) = \frac{1}{N}\sum_{i=1}^{N}\text{sgn}\, g(i\Delta t)[g(i+m)\Delta t + g((i+m+2)\Delta t) - 2g((i+m+1)\Delta t)]$$

It is clear that in the period T_0, due to the absence of correlation between $X(i\Delta t)$ and $\varepsilon(i\Delta t)$, the estimate of the cross-correlation function $R_{X\varepsilon}(\mu = 0)$ between the useful signal and the noise will be close to zero. It is also obvious that during the initiation of various defects that precede accidents, in the period T_1, as a result of the emergence of $\varepsilon_2(i\Delta t)$ ($\varepsilon(i\Delta t) = \varepsilon_1(i\Delta t) + \varepsilon_2(i\Delta t)$), the value of the estimate of the relay cross-correlation function will increase sharply due to the presence of a correlation between $X(i\Delta t)$ and $\varepsilon(i\Delta t)$. The distinctive feature of this algorithm is that during the initiation of various faults, when a correlation emerges between $X(i\Delta t)$ and $\varepsilon(i\Delta t)$, the estimates of $R^*_{X\varepsilon}(1\Delta t)$, $R^*_{X\varepsilon}(2\Delta t)$, $R^*_{X\varepsilon}(3\Delta t)$, ... expressly reflect the accident development dynamics, which makes it possible to provide reliable information on the dynamics of the development of malfunction.

2.4 Technology for Analyzing the Estimate of the Density of Distribution of Noise by Its Equivalent Samples

Let us now consider the possibility of calculating approximate values of samples of the noise $\varepsilon(i\Delta t)$, which cannot be measured directly. It is clear that if we had the values of the samples of the noise $\varepsilon(i\Delta t)$ in digital form, we could calculate the noise variance from the following known expression:

$$D_\varepsilon = \frac{1}{N}\sum_{i=1}^{N}\varepsilon^2(i\Delta t).$$

2.4 Technology for Analyzing the Estimate of the Density of ...

However, it is impossible to directly extract the samples of the noise $\varepsilon(i\Delta t)$ from the samples of the noisy signal $g(i\Delta t)$. At the same time, it is possible to calculate the noise variance D_ε from Expression (2.14). Obviously, taking this expression into account and introducing the notation

$$g^2(i\Delta t) + g(i\Delta t)g((i+2)\Delta t) - 2g(i\Delta t)g((i+1)\Delta t) = \varepsilon'(i\Delta t), \quad (2.18)$$

formula (2.18) for calculating the approximate equivalent values of the samples of the noise $\varepsilon^e(i\Delta t)$ can be represented as

$$\varepsilon^e(i\Delta t) = \operatorname{sgn} \varepsilon'(i\Delta t)$$
$$\times \sqrt{\left|g^2(i\Delta t) + g(i\Delta t)g((i+2)\Delta t) - 2g(i\Delta t)g((i+1)\Delta t)\right|}$$
$$= \operatorname{sgn} \varepsilon'(i\Delta t)\sqrt{|\varepsilon'(i\Delta t)|}, \quad (2.19)$$

where $\operatorname{sgn} \varepsilon'(i\Delta t)$ is the sign of the radicand. Numerous experiments have shown that, despite possible deviations of the approximate values of the samples of $\varepsilon^*(i\Delta t)$ from their true values $\varepsilon(i\Delta t)$ by $\varepsilon^*(i\Delta t) - \varepsilon(i\Delta t)$, the following equality takes place between the estimates of their variances:

$$P\left\{\frac{1}{N}\sum_{i=1}^{N}\varepsilon^{*2}(i\Delta t) - \frac{1}{N}\sum_{i=1}^{N}\varepsilon^2(i\Delta t) \approx 0\right\} = 1,$$

$$P\left\{\frac{1}{N}\sum_{i=1}^{N}\varepsilon^*(i\Delta t) - \frac{1}{N}\sum_{i=1}^{N}\varepsilon(i\Delta t) \approx 0\right\} = 1,$$

where P is the probability sign. This equality shows that by processing $\varepsilon^*(i\Delta t)$ it is possible to get results that would be identical to the results of the analysis of the noise $\varepsilon(i\Delta t)$.

The analysis of Expression (2.19) will be discussed in detail in the next section of the book.

The possibility of calculating the approximate values of equivalent samples of the noise $\varepsilon^e(i\Delta t)$ opens a possibility for finding the estimates of the statistical characteristics of both the noise and the useful signal from the following expressions:

$$D_\varepsilon = \frac{1}{N}\sum_{i=1}^{N}\varepsilon^2(i\Delta t) \approx \frac{1}{N}\sum_{i=1}^{N}\varepsilon^{e2}(i\Delta t),$$

$$R_{X\varepsilon}(\mu) = \frac{1}{N}\sum_{i=1}^{N}X(i\Delta t)\varepsilon(i\Delta t)$$

$$\approx \frac{1}{N}\sum_{i=1}^{N}\left[g(i\Delta t) - \operatorname{sgn}\varepsilon'(i\Delta t)\sqrt{|\varepsilon'(i\Delta t)|}\right]\operatorname{sgn}\varepsilon'(i\Delta t)\sqrt{|\varepsilon'(i\Delta t)|},$$

$$D_x = \frac{1}{N}\sum_{i=1}^{N} X^2(i\Delta t) \approx \frac{1}{N}\sum_{i=1}^{N}\left[g(i\Delta t) - \varepsilon^e(i\Delta t)\right]^2$$

$$= \frac{1}{N}\sum_{i=1}^{N}\left[g(i\Delta t) - \operatorname{sgn}\varepsilon'(i\Delta t)\sqrt{|\varepsilon'(i\Delta t)|}\right]^2.$$

With the approximate values of $\varepsilon^e(i\Delta t)$ known, we can calculate the coefficient of correlation between the useful signal and the noise from the expression [1, 2]

$$r_{x\varepsilon} = \frac{R_{X\varepsilon}(0)}{\sqrt{R_{XX}(0)R_{\varepsilon\varepsilon}(0)}} = \frac{R_{X\varepsilon}(0)}{\sqrt{D_X D_\varepsilon}}$$

$$\approx \frac{1}{N}\sum_{i=1}^{N}\Big\{\left[g(i\Delta t) - \operatorname{sgn}[\varepsilon'(i\Delta t))]\sqrt{|\varepsilon'(i\Delta t)|}\right]$$

$$\times \operatorname{sgn}[\varepsilon'(i\Delta t))]\sqrt{|\varepsilon'(i\Delta t)|}$$

$$\times \left\{\left[\frac{1}{N}\sum_{i=1}^{N}\left[g(i\Delta t) - \operatorname{sgn}[\varepsilon'(i\Delta t))]\sqrt{|\varepsilon'(i\Delta t)|}\right]^2\right]\right.$$

$$\left.\times \left[\frac{1}{N}\sum_{i=1}^{N}[\varepsilon'(i\Delta t)]\right]\right\}^{-\frac{1}{2}}.$$

It is known that in order to construct the noise distribution law, it is necessary to determine N of the corresponding values of its curve $W(\varepsilon)$ from N of the samples of the noise $\varepsilon(i\Delta t)$ over the time $T = N\Delta t$. For this purpose, we can use the approximate value of the samples of the noise analog $\varepsilon_2^a(i\Delta t)$. For the case of zero correlation between $X(i\Delta t)$ and $\varepsilon(i\Delta t)$, using the number of the samples $N_1, N_2, \ldots, N_m$ of the values $\varepsilon^e(i\Delta t) = \operatorname{sgn}\varepsilon'(i\Delta t)\sqrt{|\varepsilon'(i\Delta t)|}$ given in the range from 0 to $\varepsilon_{\max}$ at equal intervals Δt, we can construct the distribution law $W(\varepsilon)$

$$W(i\Delta t) = \operatorname{sgn}[\varepsilon'(i\Delta t)]\sqrt{|\varepsilon'(i\Delta t)|}.$$

When the coefficient of correlation equals zero, the distribution law of the noise $\varepsilon(i\Delta t)$ is constructed in the following way. The minimum value $\varepsilon_{\min}$ is specified for all samples of approximate values of the noise $\varepsilon(i\Delta t)$, and the condition $[\varepsilon_{\min} + j\Delta x] \leq \varepsilon(i\Delta t) \leq [\varepsilon_{\min} + (j+1)\Delta x]$ is verified.

To construct the distribution law $W(\varepsilon)$, the number of samples $N_0, N_1, N_2, \ldots, N_m$ at which the specified conditions are fulfilled is determined successively for the values $j = 0, j = 1, j = 2, \ldots, j = m$. It is clear that using them we can construct the curve $W[\varepsilon(i\Delta t)]$ of the distribution law of the approximate values of the samples of the noise. As the number of samples N increases, this curve tends to the distribution law of the noise, i.e., to $W(\varepsilon)$.

In conclusion, it should be noted that the resources of modern personal computers make it possible to easily implement the above described technology in practice.

References

1. Aliev TA, Rzayev AH, Guluyev GA et al (2018) Robust technology and system for management of sucker rod pumping units in oil wells. Mech Syst Signal Process 99:47–56. https://doi.org/10.1016/j.ymssp.2017.06.010
2. Aliev TA, Alizada TA, Rzayeva NE et al (2017) Noise technologies and systems for monitoring the beginning of the latent period of accidents on fixed platforms. Mech Syst Signal Process 87:111–123. https://doi.org/10.1016/j.ymssp.2016.10.014
3. Aliev TA, Rzaeva NE (2017) Techniques for determining robust estimates of correlation functions for random noisy signals. Meas Tech 60(4):343–349. https://doi.org/10.1007/s11018-017-1199-y
4. Aliev TA, Guliev GA, Pashayev FH et al (2011) Algorithms for determining the coefficient of correlation and cross-correlation function between a useful signal and noise of noisy technological parameters. Cybern Syst Anal 47(3):481–489. https://doi.org/10.1007/s10559-011-9329-z
5. Aliev TA (2001) Robust methodology for improving the conditionedness of correlation matrices. Autom Control Comput Sci 35(1):8–20
6. Aliev TA, Guluyev GA, Rzayev AH et al (2009) Correlation indicators of microchanges in technical states of control objects. Cybern Syst Anal 45(4):655–662. https://doi.org/10.1007/s10559-009-9132-2
7. Aliev TA, Aliev ER, Mastalieva DI et al (2008) Adaptive technology of the sampling of noise-corrupted signals. Autom Control Comput Sci 42(1):20–25
8. Aliev TA, Musaeva NF, Sattarova UE (2010) Robust technologies for calculating normalized correlation functions. Cybern Syst Anal 46(1):153–166. https://doi.org/10.1007/s10559-010-9194-1
9. Aliev TA, Musaeva NF, Sattarova UE (2011) Technology of calculating robust normalized correlation matrices. Cybern Syst Anal 47(1):152–165. https://doi.org/10.1007/s10559-011-9298-2
10. Aliev TA, Amirov ZA (1998) Algorithm to choose the regularization parameters for statistical identification. Autom Remote Control 59(6):859–866
11. Aliev TA, Musaeva NF (1995) Algorithm for reducing errors in estimation of a correlation function of a noisy signal. Optoelectron, Instrum Data Process 4:100–106
12. Aliev TA, Musaeva NF (1998) An algorithm for eliminating microerrors of noise in the solution of statistical dynamics problems. Autom Remote Control 59(5):679–688
13. Aliev TA, Mamedova UM (2003) Positional binary methodology for extraction of interference from noisy signals. Autom Remote Control 37(2):12–19

Chapter 3
Algorithms for Forming Correlation Matrices Equivalent to Matrices of Useful Signals in the Latent Period of Object's Emergency State

Abstract The chapter analyses the difficulties that arise in the formation of correlation matrices for input and output signals in the latent period of an emergency state of industrial facilities, showing that due to significant errors in the estimates of their elements caused by the noise correlated with the useful signal, the use of traditional methods in most cases does not ensure adequate results of solving of many applied problems. In many real industrial facilities, input and output variables are usually various physical quantities and additional errors arise during the formation of normalized correlation matrices, which also disrupts the adequacy of the obtained results. Technologies for forming correlation matrices equivalent to the correlation matrices of useful signals are proposed. This also eliminates the errors caused by the normalization of elements of equivalent normalized correlation matrices. In addition, the possibility of forming correlation matrices of noisy signals equivalent to the matrices of their useful signals by correcting the samples of the analyzed signals by means of samples of equivalent noises is considered. It is also shown that equivalent matrices, in addition to controlling the beginning and dynamics of changes in the emergency state of facilities, also improve the adequacy of the results of solving identification and management problems in various areas of industry, power engineering, transport, etc.

3.1 Difficulties in Forming the Correlation Matrices in the Latent Period of Object's Emergency State

At present, there are substantial difficulties in forming the correlation matrices of noisy input and output signals in the latent period of object's emergency state. The first prerequisite for their elimination is technologies for forming the correlation matrices that are equivalent to the matrices of the technical condition of objects during their transition to an emergency state. The urgency of this problem is due to the fact that the control of the beginning and dynamics of changes in an emergency state of objects is essential in all areas of industry, energy, and transport.

It is known that in solving this problem in the normal mode of object's operation, the impulse response $W(t)$ is determined on the basis of measurement of the input signal $X(t)$ and output signal $Y(t)$. To do this, we need to solve a system of linear algebraic equations, which in matrix form is written as follows [1–4]:

$$\vec{R}_{YX} = \vec{R}_{XX} \vec{W} \Delta\tau, \tag{3.1}$$

where $\vec{R}_{XX}$ is the square $m \times m$ symmetric matrix of the autocorrelation functions of the centered input signal $X(t)$

$$\vec{R}_{XX} = \begin{bmatrix} R_{XX}(0) & R_{XX}(\Delta\tau) & \ldots & R_{XX}[(m-1)\Delta\tau] \\ R_{XX}(\Delta\tau) & R_{XX}(0) & \ldots & R_{XX}[(m-2)\Delta\tau] \\ \ldots & \ldots & \ldots & \ldots \\ R_{XX}[(m-1)\Delta\tau] & R_{XX}[(m-2)\Delta\tau] & \ldots & R_{XX}(0) \end{bmatrix}, \tag{3.2}$$

$\vec{R}_{XX}$ is the column vector of the cross-correlation functions between the input $X(t)$ and the output $Y(t)$

$$\vec{R}_{YX} = \begin{bmatrix} R_{YX}(0) & R_{YX}(\Delta\tau) & \ldots & R_{YX}[(m-1)\Delta\tau] \end{bmatrix}^{\mathrm{T}}, \tag{3.3}$$

$\vec{W}$ is the column vector, whose elements are the ordinates of the sought-for impulsive admittance functions

$$\vec{W} = \begin{bmatrix} W(0) & W(\Delta\tau) & \ldots & W[(m-1)\Delta\tau] \end{bmatrix}^{\mathrm{T}}$$

The matrices of correlation functions (3.2), (3.3) are formed from the estimates of the useful signals $X(t)$ and $Y(t)$, which are calculated from the formulas

$$R_{XX}(\mu) = M[X(t)X(t+\mu\Delta t)] = \frac{1}{N}\sum_{i=1}^{N} X(i\Delta t)X(i\Delta t + \mu\Delta t),$$

$$n = N - k\mu,$$

$$\vec{R}_{YX} = M[Y(t)X(t+\mu\Delta t)] = \frac{1}{N}\sum_{i=1}^{N} Y(i\Delta t)X(i\Delta t + \mu\Delta t).$$

The notation here is as follows. Assume that the time interval $[0, T]$ consists of N of very small intervals Δt, i.e., $T = N\Delta t$. Assign to t discrete values multiple of Δt, i.e. $t = \mu\Delta t, \mu = \overline{0, N}$, and introduce the following notation for the estimates of the correlation functions: $R_{gg}(\mu\Delta t) = R_{gg}(\mu)$, $R_{XX}(\mu\Delta t) = R_{XX}(\mu)$, $R_{g\eta}(\mu\Delta t) = R_{g\eta}(\mu)$, $R_{XY}(\mu\Delta t) = R_{XY}(\mu)$.

Since in reality signals are distorted by noise, the noisy input $g(t)$ and output $\eta(t)$ signals of the object are the sum of the useful signals $X(t)$ and $Y(t)$ of the corresponding noise $\varepsilon(t)$ and $\varphi(t)$, i.e.,

$$g(t) = X(t) + \varepsilon(t),$$

3.1 Difficulties in Forming the Correlation Matrices ...

$$\eta(t) = Y(t) + \varphi(t).$$

Then matrix Eq. (3.1) can be written in the following form:

$$\vec{R}_{\eta g} = \vec{R}_{gg} \vec{W} \Delta \tau,$$

where

$$\vec{R}_{gg} = \begin{bmatrix} R_{gg}(0) & R_{gg}(\Delta\tau) & \dots & R_{gg}[(m-1)\Delta\tau] \\ R_{gg}(\Delta\tau) & R_{gg}(0) & \dots & R_{gg}[(m-2)\Delta\tau] \\ \dots & \dots & \dots & \dots \\ R_{gg}[(m-1)\Delta\tau] & R_{gg}[(m-2)\Delta\tau] & \dots & R_{gg}(0) \end{bmatrix}, \quad (3.4)$$

$$\vec{R}_{\eta g} = \begin{bmatrix} R_{\eta g}(0) & R_{\eta g}(\Delta\tau) & \dots & R_{\eta g}[(m-1)\Delta\tau] \end{bmatrix}^{\mathrm{T}}. \quad (3.5)$$

Here

$$\left. \begin{array}{l} R_{gg}(\mu) = M[g(t)g(t+\mu\Delta t)] \\ R_{\eta g}(\mu) = M[\eta(t)g(t+\mu\Delta t)] \end{array} \right\}. \quad (3.6)$$

In practice, correlation matrices (3.4) and (3.5) are formed from the estimates $R_{gg}(\mu)$ and $R_{\eta g}(\mu)$ of the correlation functions of the noisy signals $g(t)$ and $\eta(t)$. However, in this case, the following obvious inequalities arise:

$$\left. \begin{array}{l} \vec{R}_{XX} \neq \vec{R}_{gg} \\ \vec{R}_{YX} \neq \vec{R}_{\eta g} \end{array} \right\}.$$

As a result, in many cases it is impossible to ensure adequate identification of object's technical condition even under normal operation conditions and if the above-mentioned classical conditions are fulfilled.

At the same time, in many real control objects, noisy signals are measured by various sensors, in which the input and output signals represent different physical quantities. In these cases, the estimates of the correlation functions of the signals $X(t)$ and $Y(t)$ are reduced to dimensionless quantities, and their normalized auto- and cross-correlation functions are calculated from the known formulas [1–4]

$$\left. \begin{array}{l} r_{XX}(\mu) = \frac{R_{XX}(\mu)}{D_X} \\ r_{YX}(\mu) = \frac{R_{YX}(\mu)}{\sqrt{D_X D_Y}} \end{array} \right\}, \quad (3.7)$$

where $D_X = R_{XX}(0)$ and $D_Y = R_{YY}(0)$ are the variances of the useful signals $X(t)$ and $Y(t)$, respectively.

The normalized correlation matrices of the useful signals have the following form:

$$\vec{r}_{XX} = \begin{bmatrix} 1 & r_{XX}(\Delta\tau) & \ldots & r_{XX}[(m-1)\Delta\tau] \\ r_{XX}(\Delta\tau) & 1 & \ldots & r_{XX}[(m-2)\Delta\tau] \\ \ldots & \ldots & \ldots & \ldots \\ r_{XX}[(m-1)\Delta\tau] & r_{XX}[(m-2)\Delta\tau] & \ldots & 1 \end{bmatrix}, \quad (3.8)$$

$$\vec{r}_{YX} = \begin{bmatrix} r_{YX}(0) & r_{YX}(\Delta\tau) & \ldots & r_{YX}[(m-1)\Delta\tau] \end{bmatrix}^{\mathrm{T}}. \quad (3.9)$$

The normalized auto- and cross-correlation functions $r_{gg}(\mu)$ and $r_{\eta g}(\mu)$ of the noisy signals $g(t)$ and $\eta(t)$, which are the sum of the random useful signals $X(t)$ and $Y(t)$ and the corresponding noises $\varepsilon(t)$ and $\varphi(t)$, respectively, are calculated from the formulas [1, 2]:

$$\left. \begin{array}{l} r_{gg}(\mu) = R_{gg}(\mu)/D_g \\ r_{\eta g}(\mu) = R_{\eta g}(\mu)/\sqrt{D_g D_\eta} \end{array} \right\}. \quad (3.10)$$

In this case, the corresponding normalized correlation matrices of the noisy signals $g(i\Delta t)$ and $\eta(i\Delta t)$ are written in the following form:

$$\vec{r}_{gg}(\mu) = \begin{Vmatrix} \frac{R_{gg}(0)}{D_g} & \frac{R_{gg}(\Delta t)}{D_g} & \ldots & \frac{R_{gg}[(N-1)\Delta t]}{D_g} \\ \frac{R_{gg}(\Delta t)}{D_g} & \frac{R_{gg}(0)}{D_g} & \ldots & \frac{R_{gg}[(N-2)\Delta t]}{D_g} \\ \ldots & \ldots & \ldots & \ldots \\ \frac{R_{gg}[(N-1)\Delta t]}{D_g} & \frac{R_{gg}[(N-2)\Delta t]}{D_g} & \ldots & \frac{R_{gg}(0)}{D_g} \end{Vmatrix}, \quad (3.11)$$

$$\vec{r}_{g\eta}(\mu) = \begin{bmatrix} \frac{R_{g\eta}(0)}{\sqrt{D_g D_\eta}} & \frac{R_{g\eta}(\Delta t)}{\sqrt{D_g D_\eta}} & \ldots & \frac{R_{g\eta}[(N-1)\Delta t]}{\sqrt{D_g D_\eta}} \end{bmatrix}^{\mathrm{T}}. \quad (3.12)$$

Comparing matrices (3.8) and (3.11), it is easy to see that the following inequalities take place:

$$\left. \begin{array}{l} \vec{r}_{gg}(\mu) \neq \vec{r}_{XX}(\mu) \\ \vec{r}_{g\eta}(\mu) \neq \vec{r}_{XY}(\mu) \end{array} \right\}. \quad (3.13)$$

It follows from inequalities (3.7) and (3.13) that correlation matrices (3.4), (3.5) and (3.11), (3.12) differ from original correlation matrices (3.2), (3.3) and (3.8), (3.9). For this exact reason, in practice, it is often impossible to ensure adequate control of object's technical condition based on these matrices even in the normal operation mode. In this regard, in order to successfully solve the problem of controlling the beginning and development dynamics of accidents at facilities, it is necessary to create a technology for forming the equivalent correlation matrices $\vec{R}^e_{gg}(\mu)$, $\vec{R}^e_{g\eta}(\mu)$ and $\vec{r}^e_{gg}(\mu)$, $\vec{r}^e_{g\eta}(\mu)$ that ensure the fulfillment of the equalities

$$\left. \begin{array}{l} \vec{R}^e_{gg}(\mu) \approx \vec{R}_{XX}(\mu) \\ \vec{R}^e_{g\eta}(\mu) \approx \vec{R}_{XY}(\mu) \\ \vec{r}^e_{gg}(\mu) \approx \vec{r}_{XX}(\mu) \\ \vec{r}^e_{g\eta}(\mu) \approx \vec{r}_{XY}(\mu) \end{array} \right\}.$$

3.2 Technologies for Forming the Equivalent Correlation Matrices for an Object in the Normal Operating Mode

The research in [1–9] has demonstrated that the conditions of stationarity and normality of distribution law hold for the input and output noisy signals of many control objects. For this reason, there is no correlation between the useful signals $X(i\Delta t)$ and $Y(i\Delta t)$ and the noises $\varepsilon(i\Delta t)$, $\varphi(i\Delta t)$, i.e.,

$$\left.\begin{array}{l}\frac{1}{N}\sum_{i=1}^{N}X(i\Delta t)\varepsilon((i+\mu)\Delta t)=0\\ \frac{1}{N}\sum_{i=1}^{N}Y(i\Delta t)\varphi((i+\mu)\Delta t)=0\\ \frac{1}{N}\sum_{i=1}^{N}\varepsilon(i\Delta t)\varepsilon((i+\mu)\Delta t)=0\\ \frac{1}{N}\sum_{i=1}^{N}\varphi(i\Delta t)\varphi((i+\mu)\Delta t)=0\\ \frac{1}{N}\sum_{i=1}^{N}\varepsilon(i\Delta t)\varphi((i+1)\Delta t)=0\end{array}\right\} \quad (3.14)$$

and expression (3.6) for calculating the estimates of the auto- and cross-correlation functions can be represented in the form

$$R_{gg}(\mu) = \frac{1}{N}\sum_{i=1}^{N}g(i\Delta t)g((i+\mu)\Delta t)$$

$$= \frac{1}{N}\sum_{i=1}^{N}(X(i\Delta t)+\varepsilon(i\Delta t))(X((i+\mu)\Delta t)+\varepsilon((i+\mu)\Delta t))$$

$$\approx \begin{cases} R_{XX}(0) + D_\varepsilon & \text{when } \mu = 0 \\ R_{XX}(\mu) & \text{when } \mu \neq 0 \end{cases} \quad (3.15)$$

$$R_{g\eta}(\mu) = \frac{1}{N}\sum_{i=1}^{N}g(i\Delta t)\eta((i+\mu)\Delta t)$$

$$= \frac{1}{N}\sum_{i=1}^{N}(X(i\Delta t)+\varepsilon(i\Delta t))(Y((i+\mu)\Delta t)$$

$$+\varphi((i+\mu)\Delta t)) \approx R_{XY}(\mu). \quad (3.16)$$

Therefore, considering that according to expression $R_{XX}(0) = R_{gg}(0) - D_\varepsilon$ from (3.15), the correlation matrix $\vec{R}_{gg}(\mu)$ from (3.4) equivalent to the matrix in the formula from (3.2) can be represented as follows:

$$\vec{R}^e_{gg}(\mu) = \begin{Vmatrix} R_{gg}(0) - D_\varepsilon & R_{gg}(\Delta t) & \ldots & R_{gg}[(N-1)\Delta t] \\ R_{gg}(\Delta t) & R_{gg}(0) - D_\varepsilon & \ldots & R_{gg}[(N-2)\Delta t] \\ \ldots & \ldots & \ldots & \ldots \\ R_{gg}[(N-1)\Delta t] & R_{gg}[(N-2)\Delta t] & \ldots & R_{gg}(0) - D_\varepsilon \end{Vmatrix}$$

$$\approx \begin{Vmatrix} R_{XX}(0) & R_{XX}(\Delta t) & \ldots & R_{XX}[(N-1)\Delta t] \\ R_{XX}(\Delta t) & R_{XX}(0) & \ldots & R_{XX}[(N-2)\Delta t] \\ \ldots & \ldots & \ldots & \ldots \\ R_{XX}[(N-1)\Delta t] & R_{XX}[(N-2)\Delta t] & \ldots & R_{XX}(0) \end{Vmatrix}. \quad (3.17)$$

Based on expression (3.16), taking into account that in the absence of correlation between $X(i\Delta t)$, $\varepsilon(i\Delta t)$ and $Y(i\Delta t)$, $\varphi(i\Delta t)$ the following equality takes place:

$$R_{g\eta}(0) = R_{XY}(0), \ R_{g\eta}(\Delta t) = R_{XY}(\Delta t), \ldots,$$
$$R_{g\eta}((N-1)\Delta t) = R_{XY}((N-1)\Delta t),$$

matrix (3.5) can be represented as equivalent correlation matrix (3.3)

$$\vec{R}^e_{g\eta}(\mu) \approx \left[R_{XY}(0) \ R_{XY}(\Delta t) \ \ldots \ R_{XY}[(N-1)\Delta t] \right]^T = \vec{R}_{XY}(\mu). \quad (3.18)$$

Our experimental research has also demonstrated that for objects fulfilling conditions (3.14), by calculating the estimates of the elements of $R_{g\eta}(\mu)$ from expression (3.16), it is possible to form the equivalent matrices $\vec{R}^e_{g\eta}(\mu)$ from (3.18), whose elements match the elements of the correlation matrix $\vec{R}_{XY}(\mu)$ of the useful signals $X(i\Delta t)$ and $Y(i\Delta t)$ from (3.3).

However, the correlation matrices $\vec{R}_{gg}(\mu)$ from (3.4) and (3.17) of the noisy input signal $g(i\Delta t)$ differs from the correlation matrix $\vec{R}_{XX}(\mu)$ from (3.2) of the useful signal $X(i\Delta t)$ in its diagonal elements, which are the sum of the estimates of the correlation function $R_{XX}(0)$ of the useful signals and the noise variance D_ε. Therefore, in practice, to form the equivalent matrix $\vec{R}^e_{gg}(\mu)$ from (3.17), it is necessary to correct the estimates $R_{gg}(0)$ of the diagonal elements of the matrix (3.4).

Therefore, the formation of matrix (3.17) equivalent to matrix (3.2) comes down to eliminating the errors of the noise D_ε from the diagonal elements of matrix (3.4).

In addition, the normalization of the estimates of the correlation functions also requires calculating D_ε. This is because in this case taking into account expressions (3.15), Formulas (3.10) can be transformed to a form similar to (3.7), i.e.,

$$r_{gg}(\mu \neq 0) = \frac{R_{gg}(\mu \neq 0)}{D_g - D_\varepsilon} \approx \frac{R_{XX}(\mu \neq 0)}{D_X}, \quad (3.19)$$

$$r_{g\eta}(\mu) = \frac{R_{g\eta}(\mu)}{\sqrt{(D_g - D_\varepsilon)(D_\eta - D_\varphi)}} \approx \frac{R_{XY}(\mu)}{\sqrt{D_X D_Y}}. \quad (3.20)$$

3.2 Technologies for Forming the Equivalent Correlation Matrices ...

In this case, the normalized correlation matrix of the noisy signals $g(i\Delta t)$ from (3.11) equivalent to the matrix (3.8) can be represented as

$$\vec{r}^{\,e}_{gg}(\mu) = \begin{Vmatrix} 1 & \frac{R_{gg}(\Delta t)}{D_g-D_\varepsilon} & \cdots & \frac{R_{gg}[(N-1)\Delta t]}{D_g-D_\varepsilon} \\ \frac{R_{gg}(\Delta t)}{D_g-D_\varepsilon} & 1 & \cdots & \frac{R_{gg}[(N-2)\Delta t]}{D_g-D_\varepsilon} \\ \cdots & \cdots & \cdots & \cdots \\ \frac{R_{gg}[(N-1)\Delta t]}{D_g-D_\varepsilon} & \frac{R_{gg}[(N-2)\Delta t]}{D_g-D_\varepsilon} & \cdots & 1 \end{Vmatrix}$$

$$\approx \begin{Vmatrix} 1 & \frac{R_{XX}(\Delta t)}{D_X} & \cdots & \frac{R_{XX}(N-1)\Delta t}{D_X} \\ \frac{R_{XX}(\Delta t)}{D_X} & 1 & \cdots & \frac{R_{XX}(N-2)\Delta t}{D_X} \\ \cdots & \cdots & \cdots & \cdots \\ \frac{R_{XX}(N-1)\Delta t}{D_X} & \frac{R_{gg}[(N-2)\Delta t] \approx R_{XX}(N-2)\Delta t}{D_X} & \cdots & 1 \end{Vmatrix}. \quad (3.21)$$

Similarly, it is also possible to form from (3.12) a matrix equivalent to the matrix of the normalized cross-correlation functions (3.9), i.e.,

$$\vec{r}^{\,e}_{g\eta}(\mu) = \left[\frac{R_{g\eta}(0)}{\sqrt{(D_g-D_\varepsilon)(D_\eta-D_\varphi)}} \quad \frac{R_{g\eta}(\Delta t)}{\sqrt{(D_g-D_\varepsilon)(D_\eta-D_\varphi)}} \quad \cdots \quad \frac{R_{g\eta}[(N-1)\Delta t]}{\sqrt{(D_g-D_\varepsilon)(D_\eta-D_\varphi)}} \right]^T$$

$$\approx \left[\frac{R_{XY}(0)}{\sqrt{D_X D_Y}} \quad \frac{R_{XY}(\Delta t)}{\sqrt{D_X D_Y}} \quad \cdots \quad \frac{R_{XY}[(N-1)\Delta t]}{\sqrt{D_X D_Y}} \right]^T. \quad (3.22)$$

Thus, according to expressions (3.15)–(3.22), by eliminating the error of noise D_ε and D_φ, it is possible to form from the corresponding elements of the matrices of noisy signals correlation matrices (3.17), (3.18), (3.21), and (3.22), which will be equivalent to corresponding matrices (3.1), (3.3), (3.8), and (3.9) of the useful signals. However, to do this, we need to calculate the estimates of the variances D_ε and D_φ of the noise of the noisy signals $g(i\Delta t)$ and $\eta(i\Delta t)$. Our research has shown that for this purpose it is expedient to use expressions (2.12) from Chap. 2.

This shows that after the correction of the corresponding elements of matrices (3.17), (3.18) and (3.21), (3.22) by the estimates of the noise variance D_ε, they can be regarded as equivalent to matrices (3.2), (3.3) and (3.8), (3.9), respectively.

For instance, after subtracting the estimate D_ε from the estimate of the diagonal elements $R_{gg}(0)$ of matrix (3.17), the estimates of all its elements become equal to the corresponding estimates of matrix (3.2) of the useful signals.

Thus, the possibility of calculating the estimates of the noise variances D_ε and D_φ of the noisy signals $g(i\Delta t)$ and $\eta(i\Delta t)$ allows us to form the equivalent matrices $\vec{R}^{\,e}_{gg}(\mu)$, $\vec{R}^{\,e}_{g\eta}(\mu)$ and $\vec{r}^{\,e}_{gg}(\mu)$, $\vec{r}^{\,e}_{g\eta}(\mu)$, in which the estimates of the elements are respectively equal to the estimates of the elements of the matrices $\vec{R}_{XX}(\mu)$, $\vec{R}_{XY}(\mu)$ and $\vec{r}_{XX}(\mu)$, $\vec{r}_{XY}(\mu)$ of the useful signals. Therefore, in the absence of a correlation between $g(i\Delta t)$ and $\varepsilon(i\Delta t)$, $Y(i\Delta t)$ and $\varphi(i\Delta t)$, we can assume that the following equalities take place between the matrices of the useful signals and the equivalent matrices of noisy signals, i.e.,

$$\left.\begin{array}{l}\vec{R}^e_{gg}(\mu) \approx \vec{R}_{XX}(\mu) \\ \vec{R}^e_{g\eta}(\mu) \approx \vec{R}_{XY}(\mu) \\ \vec{r}^e_{gg}(\mu) \approx \vec{r}_{XX}(\mu) \\ \vec{r}^e_{g\eta}(\mu) \approx \vec{r}_{XY}(\mu)\end{array}\right\}$$

3.3 Technology for Forming the Equivalent Correlation Matrix in the Latent Period of Object's Emergency State

As noted earlier, it is characteristic of many real-life industrial facilities during operation to develop various defects, such as wear and tear, microcracks, carbon deposition, fatigue strain, etc. They usually affect the signals received from the corresponding sensors as the noises $\varepsilon_2(i\Delta t)$ and $\varphi_2(i\Delta t)$ that in most cases correlate with the useful signals $X(i\Delta t)$ and $Y(i\Delta t)$ [3, 4]. The sum noise in such cases forms from the noise $\varepsilon_1(i\Delta t)$, which is caused by the external factors, and the noise $\varepsilon_2(i\Delta t)$ that emerges as a result of the initiation of various defects. The variance of the noisy signal, in that case, takes the following form:

$$D_g = R_{gg}(0) = \frac{1}{N}\sum_{i=1}^{N} g^2(i\Delta t)$$

$$= \frac{1}{N}\sum_{i=1}^{N} X^2(i\Delta t) + 2\frac{1}{N}\sum_{i=1}^{N} X(i\Delta t)\varepsilon(i\Delta t) + \frac{1}{N}\sum_{i=1}^{N} \varepsilon(i\Delta t)\varepsilon(i\Delta t)$$

$$= R_{XX}(0) + 2R_{\varepsilon X}(0) + D_{\varepsilon\varepsilon}.$$

Therefore, when the sum noise

$$\varepsilon(i\Delta t) = \varepsilon_1(i\Delta t) + \varepsilon_2(i\Delta t)$$

has a correlation with the useful signal $X(i\Delta t)$, its variance D_ε is calculated from the expression

$$D_\varepsilon \approx 2R_{X\varepsilon}(0) + D_{\varepsilon\varepsilon},$$

where

$$R_{X\varepsilon}(0) = \frac{1}{N}\sum_{i=1}^{N} X(i\Delta t)\varepsilon(i\Delta t), \quad D_{\varepsilon\varepsilon} = \frac{1}{N}\sum_{i=1}^{N} \varepsilon(i\Delta t)\varepsilon(i\Delta t).$$

3.3 Technology for Forming the Equivalent Correlation Matrix ...

In this case, the formula for calculating the estimate of $R_{gg}(\mu)$ can be represented as follows:

$$R_{gg}(\mu) = \frac{1}{N}\sum_{i=1}^{N} g(i\Delta t)g((i+\mu)\Delta t)$$

$$= \frac{1}{N}\sum_{i=1}^{N}[X(i\Delta t)+\varepsilon(i\Delta t)][X((i+\mu)\Delta t)+\varepsilon((i+\mu)\Delta t)]$$

$$= \frac{1}{N}\sum_{i=1}^{N}[X(i\Delta t)X((i+\mu)\Delta t)+\varepsilon(i\Delta t)X((i+\mu)\Delta t)$$

$$+ X(i\Delta t)\varepsilon((i+\mu)\Delta t)$$

$$+ \varepsilon(i\Delta t)\varepsilon((i+\mu)\Delta t)] = R_{XX}(\mu) + R_{\varepsilon X}(\mu) + R_{X\varepsilon}(\mu) + R_{\varepsilon\varepsilon}(\mu)$$

$$\approx \begin{cases} R_{XX}(0) + 2R_{X\varepsilon}(0) + D_{\varepsilon\varepsilon} & \text{when } \mu = 0 \\ R_{XX}(\mu) + 2R_{X\varepsilon}(\mu) & \text{when } \mu \neq 0 \end{cases}$$

According to our experimental research, with defect development dynamics in real-life industrial facilities, a correlation between $X(i\Delta t)$ and $\varepsilon(i\Delta t)$ often takes place even during several sampling intervals, i.e., when $\mu = \Delta t, 2\Delta t, 3\Delta t, \ldots$ [3, 4]. Therefore, the formation of equivalent matrices requires developing technology for calculating the estimates of $R_{X\varepsilon}(0), R_{X\varepsilon}(\Delta t), R_{X\varepsilon}(2\Delta t), R_{X\varepsilon}(3\Delta t)\ldots$ of the cross-correlation functions between $X(i\Delta t)$ and $\varepsilon(i\Delta t)$. In this case, by compensating for the errors of the elements $R_{gg} = (0), R_{gg}(\Delta t), R_{gg}(2\Delta t), R_{gg}(3\Delta t),\ldots$ in the corresponding rows and columns of correlation matrices (3.17) and (3.21) one can ensure that they are equivalent to the matrices of the useful signals (3.2) and (3.8). Therefore, to ensure that correlation matrices (3.4) are equivalent to matrices (3.2) of the useful signals, we need to subtract the values of $2R_{X\varepsilon}(0)$ and $D_{\varepsilon\varepsilon}$ from the estimates of $R_{gg}(0)$, and the value of $2R_{X\varepsilon}(\mu)$ from the estimates of $R_{gg}(\mu)$

$$\vec{R}_{gg}^{e}(\mu) \approx \begin{Vmatrix} R_{XX}(0) & R_{XX}(\Delta t) & \ldots & R_{XX}[(N-1)\Delta t] \\ R_{XX}(\Delta t) & R_{XX}(0) & \ldots & R_{XX}[(N-2)\Delta t] \\ \ldots & \ldots & \ldots & \ldots \\ R_{XX}[(N-1)\Delta t] & R_{XX}[(N-2)\Delta t] & \ldots & R_{XX}(0) \end{Vmatrix}$$

$$\approx \vec{R}_{XX}(\mu).$$

Therefore, in order to form a matrix equivalent to matrix (3.8) of the normalized correlation function of the useful signal $X(i\Delta t)$ from (3.11), it becomes necessary to calculate the estimates of $R_{gg}(0), R_{gg}(\Delta t), \ldots, R_{gg}((N-1)\Delta t)$ and to divide

$$R_{gg}(0) - 2R_{X\varepsilon}(0) - D_{\varepsilon\varepsilon} \approx R_{XX}(0)$$
$$R_{gg}(\Delta t) - 2R_{X\varepsilon}(\Delta t) \approx R_{XX}(\Delta t)$$
$$\ldots$$
$$R_{gg}[(N-1)\Delta t] - 2R_{X\varepsilon}[(N-1)\Delta t] \approx R_{XX}[(N-1)\Delta t]$$

by D_X after the correction. However, in traditional technologies, these estimates are usually divided by the variance D_g of the sum signal. This is acceptable in cases when the equality $D_g - D_X \approx 0$ holds. At the same time, for real noisy signals this difference, i.e., $D_g - D_X = D_\varepsilon$, is a significant value. Sometimes the equality $D_\varepsilon \approx (0.1 - 0.25)D_g$ takes place. Therefore, in these cases, a normalization by traditional technologies introduces an additional error. Because of this, to ensure the equivalence of the matrices, it is necessary to divide the numerator of the elements of matrix (3.21) by the difference $D_g - D_\varepsilon = D_X$. Due to this, the matrix

$$\vec{r}_{gg}^{\,e}(\mu) \approx \begin{Vmatrix} 1 & \frac{R_{XX}(\Delta t)}{D_X} & \cdots & \frac{R_{XX}[(N-1)\Delta t]}{D_X} \\ \frac{R_{XX}(\Delta t)}{D_X} & 1 & \cdots & \frac{R_{XX}[(N-2)\Delta t]}{D_X} \\ \cdots & \cdots & \cdots & \cdots \\ \frac{R_{XX}[(N-1)\Delta t]}{D_X} & \frac{R_{XX}[(N-2)\Delta t]}{D_X} & \cdots & 1 \end{Vmatrix}$$
$$\approx \vec{r}_{XX}(\mu).$$

can be regarded as equivalent to matrix (3.8).

Thus, to form the required matrix, it is necessary to calculate the estimates of $R_{X\varepsilon}(0)$, $R_{X\varepsilon}(\Delta t)$, $R_{X\varepsilon}(2\Delta t)$, etc., of the noisy signal $g(i\Delta t)$ that can be found from the generalized expression

$$R_{X\varepsilon}(m\Delta t) \approx \frac{1}{2} R'_{X\varepsilon}(m\Delta t) \approx \frac{1}{2N} \sum_{i=1}^{N} [g(i\Delta t)g((i+m)\Delta t)$$
$$- 2g(i\Delta t)g((i+m+1)\Delta t)$$
$$+ g(i\Delta t)g((i+m+2)\Delta t)],$$

when $m = 1, 2, 3, \ldots$

Our experimental analysis of the noisy signals received at seismic-acoustic stations compressor stations, as well as fixed offshore platforms and oil and gas extraction and refining facilities [1–4] has demonstrated that due to accident development dynamics a correlation often takes place between $X(i\Delta t)$ and $\varepsilon(i\Delta t)$ at different time shifts. The maximum time shift during those experiments did not exceed $\mu = 10\Delta t$, i.e., the correlation disappeared at $\mu = 10\Delta t$. The obtained estimates can be used for correcting the results obtained by traditional technologies.

The effectiveness of the developed technology was verified through thorough long-term experiments over an extended period of time during the operation of compressor units and oil and gas extracting facilities. With these and many other objects, it becomes necessary to control the latent period of their transi-

3.3 Technology for Forming the Equivalent Correlation Matrix ...

tion to an emergency state during their operation in real time. In these cases, the identification of object's technical condition by solving the matrix equation takes a long time. Therefore, it turned out to be reasonable to control the beginning of the transition of objects from the normal state to an emergency state by means of a set of informative attributes formed from the estimates of the noise characteristics of noisy signals $g_1(i\Delta t), g_2(i\Delta t), g_3(i\Delta t), \ldots, g_m(i\Delta t)$, such as $D_{\varepsilon_i}, \ldots, R_{X_i\varepsilon_i}(0), R_{X_i\varepsilon_i}(\Delta t), R_{X_i\varepsilon_i}(2\Delta t), \ldots$, where $i = 1, 2, 3, \ldots, m$.

Our experimental studies have shown that the equivalent correlation matrices can also be formed by extracting approximate values of the samples of the noise $\varepsilon^*(i\Delta t)$ from the sum signal $g(i\Delta t)$. In this case, using the approximate values of the samples of the useful signal $X^*(i\Delta t)$, it is possible to form correlation matrices equivalent to the matrices of the useful signals $X(i\Delta t)$. To this end, the approximate samples of the useful signal $X^*(i\Delta t)$ are calculated from the formula

$$X^*(i\Delta t) \approx g(i\Delta t) - \varepsilon^*(i\Delta t) = g(i\Delta t) - \operatorname{sgn}\varepsilon'(i\Delta t)$$
$$\times \sqrt{|g^2(i\Delta t) + g(i\Delta t)g((i+2)\Delta t) - 2g(i\Delta t)g((i+1)\Delta t)|}.$$

Then, from these approximate values of the samples of the useful signal $X^*(i\Delta t)$ using the known technology, the estimates of the correlation functions $R^*_{XX}(\mu)$ and $R^*_{XY}(\mu)$ are calculated. Due to this, it is possible, by extracting the samples of the noise from the noisy signal, to form the equivalent correlation matrices $R^*_{XX}(\mu)$ and $R^*_{XX}(0)$ using the approximate values of the samples of the useful signal $X^*(i\Delta t)$. In doing that, we take into account that, in spite of certain errors in the samples $X_i^*(i\Delta t)$ compared with the samples of the useful signals $X(i\Delta t)$, the following equality holds if the observation time T is long enough:

$$\left.\begin{array}{l}
P[X_1(i\Delta t) \geq X_1^*(i\Delta t)] = P[X_1(i\Delta t) \leq X_1^*(i\Delta t)] \\
P[X_2(i\Delta t) \geq X_2^*(i\Delta t)] = P[X_2(i\Delta t) \leq X_2^*(i\Delta t)] \\
\ldots \\
P[X_n(i\Delta t) \geq X_n^*(i\Delta t)] = P[X_n(i\Delta t) \leq X_n^*(i\Delta t)] \\
P[X_1((i+\mu)\Delta t) \geq X_1^*((i+\mu)\Delta t)] = P[X_1((i+\mu)\Delta t) \leq X_1^*((i+\mu)\Delta t)] \\
P[X_2((i+\mu)\Delta t) \geq X_2^*((i+\mu)\Delta t)] = P[X_2((i+\mu)\Delta t) \leq X_2^*((i+\mu)\Delta t)] \\
\ldots \\
P[X_n((i+\mu)\Delta t) \geq X_n^*((i+\mu)\Delta t)] = P[X_n((i+\mu)\Delta t) \leq X_n^*((i+\mu)\Delta t)]
\end{array}\right\}$$

Due to this, numerous computational experiments have confirmed the validity of the equality

$$\vec{R}_{xx}(\mu) \approx R^*_{xx}(\mu), \quad \vec{R}_{xx}(0) \approx \vec{R}^*_{xx}(0),$$

$$\vec{R}_{xx}(\mu) = \begin{vmatrix} M[X_1(i\Delta t)X_1((i+\mu)\Delta t)] & M[X_1(i\Delta t)X_2((i+\mu)\Delta t)] & \ldots & M[X_1(i\Delta t)X_n((i+\mu)\Delta t)] \\ M[X_2(i\Delta t)X_1((i+\mu)\Delta t)] & M[X_2(i\Delta t)X_2((i+\mu)\Delta t)] & \ldots & M[X_2(i\Delta t)X_n((i+\mu)\Delta t)] \\ \ldots & \ldots & \ldots & \ldots \\ M[X_n(i\Delta t)X_1((i+\mu)\Delta t)] & M[X_n(i\Delta t)X_2((i+\mu)\Delta t)] & \ldots & M[X_n(i\Delta t)X_n((i+\mu)\Delta t)] \end{vmatrix}$$

$$\vec{R}_{xx}^*(\mu) = \begin{vmatrix} M\left[X_1^*(i\Delta t)X_1^*((i+\mu)\Delta t)\right] & M[X_1^*(i\Delta t)X_2^*((i+\mu)\Delta t)] & \ldots & M[X_1^*(i\Delta t)X_n^*((i+\mu)\Delta t)] \\ M\left[X_2^*(i\Delta t)X_1^*((i+\mu)\Delta t)\right] & M[X_2^*(i\Delta t)X_2^*((i+\mu)\Delta t)] & \ldots & M[X_2^*(i\Delta t)X_n^*((i+\mu)\Delta t)] \\ \ldots & \ldots & \ldots & \ldots \\ M\left[X_n^*(i\Delta t)X_1^*((i+\mu)\Delta t)\right] & M[X_n^*(i\Delta t)X_2^*((i+\mu)\Delta t)] & \ldots & M[X_n^*(i\Delta t)X_n^*((i+\mu)\Delta t)] \end{vmatrix}$$

Thus, we can assume that the correlation matrices $\vec{R}_{xx}^*(\mu)$ and $\vec{R}_{xx}^*(0)$ formed from the estimates calculated from the approximate values of samples of the useful signals $X^*(i\Delta t)$ are equivalent to the correlation matrices $\vec{R}_{xx}(\mu)$ and $\vec{R}_{xx}(0)$ of the useful signals.

Similar arguments can also be made for the column vectors $\vec{R}_{xy}(\mu)$ and $\vec{R}_{xy}^*(\mu)$, $\vec{R}_{xy}(0)$ and $\vec{R}_{xy}^*(0)$.

It follows that in the considered variant the solution to the problem of control of object's technical condition in the latent period of its emergency state can be reduced to solving the system of matrix equations

$$\vec{R}_{xy}^*(\mu) = \vec{R}_{xx}^*(\mu)\vec{W}(\mu),$$
$$\vec{R}_{xy}^*(0) = \vec{R}_{xx}^*(0)\vec{B}^*.$$

This opens up opportunities for solving a wide range of control and identification problems in various fields of technology that require forming the correlation matrices $R_{xx}^*(\mu)$, $R_{xx}^*(0)$, $R_{xy}^*(\mu)$ and $R_{xy}^*(0)$ equivalent to the correlation matrices of useful signals.

References

1. Aliev T (2007) Digital noise monitoring of defect origin. Springer, Boston. https://doi.org/10.1007/978-0-387-71754-8
2. Aliev T (2003) Robust technology with analysis of interference in signal processing. Kluwer Academic/Plenum Publishers, New York. https://doi.org/10.1007/978-1-4615-0093-3
3. Aliev TA, Rzayev AH, Guluyev GA et al (2018) Robust technology and system for management of sucker rod pumping units in oil wells. Mech Syst Signal Process 99:47–56. https://doi.org/10.1016/j.ymssp.2017.06.010
4. Aliev TA, Rzayeva NE, Sattarova UE et al (2017) Robust correlation technology for online monitoring of changes in the state of the heart by means of laptops and smartphones. Biomed Signal Process Control 31:44–51. https://doi.org/10.1016/j.bspc.2016.06.015
5. Aliev TA, Musaeva NF, Suleymanova MT et al (2016) Technology for calculating the parameters of the density function of the normal distribution of the useful component in a noisy process. J Autom Inf Sci 48(4):39–55. https://doi.org/10.1615/JAutomatInfScien.v48.i4.50
6. Aliev TA, Musaeva NF, Suleymanova MT (2017) Density function of noise distribution as an indicator for identifying the degree of fault growth in sucker rod pumping unit (SRPU). J Autom Inf Sci 49(4):1–11. https://doi.org/10.1615/JAutomatInfScien.v49.i4.10
7. Aliyev TA, Musayeva NF (1996) Statistical identification with error balancing. Journal of Computer and Systems Sciences International 34(5):119–124

8. Aliev TA, Musaeva NF (1997) Algorithms for determining a dispersion and errors caused by random signal interferences. Optoelectron Instrum Data Process 3:74–86
9. Aliev TA, Musaeva NF, Gazizade BI (2017) Algorithms for constructing a model of the noisy process by correcting the law of its distribution. J Autom Inf Sci 49(9):61–75. https://doi.org/10.1615/JAutomatInfScien.v49.i9.50

Chapter 4
Spectral Technology for Noise Control of the Beginning and Development Dynamics of the Latent Period of Accidents

Abstract The causes of additional errors in the estimates of the spectral characteristics of noisy signals due to the noise emerging in the latent period of facility's emergency state are analyzed. Algorithms and technologies for replacing non-measurable samples of the noise with their approximate equivalent values are proposed, showing the possibility of using them to calculate the estimates of the spectral characteristics of the noise both for the case when a correlation is present between the useful signal and the noise and for the case when there is no such correlation. It is shown that the use of the obtained estimates of the spectral characteristics of the noise makes it possible to increase the reliability of the results of traditional technologies of spectral analysis of noisy signals. It is also shown that using the algorithms for calculating equivalent samples of the noise, it is possible to create a technology of relay spectral noise analysis and a technology of sign spectral noise analysis, which are easy to implement in terms of hardware in control systems. According to the results of the experiments for the control of the beginning of the latent period of transition into an emergency state for sucker rod pumping units, compressor stations, drilling rigs, etc., the application of these technologies is effective primarily for facilities with a cyclic mode of operation.

4.1 Algorithms and Technologies for Calculating the Errors in the Estimates of Spectral Characteristics in the Latent Period of Object's Emergency State

It is known [1] that the estimates a_n and b_n of the spectral characteristics of the noisy signals $g(t)$ during object's operation in the normal mode are calculated from the formulas

$$a_n = \frac{2}{T} \int_0^T [x(t) + \varepsilon(t)] \cos n\omega t \, dt$$

$$= \frac{2}{T} \int_0^T [x(t)\cos n\omega t + \varepsilon(t)\cos n\omega t]\,dt, \tag{4.1}$$

$$b_n = \frac{2}{T} \int_0^T [x(t) + \varepsilon(t)]\sin n\omega t\,dt$$

$$= \frac{2}{T} \int_0^T [x(t)\sin n\omega t + \varepsilon(t)\sin n\omega t]\,dt. \tag{4.2}$$

Here, when the classical conditions are fulfilled, such as stationarity, the absence of correlation between $x(t)$ and $\varepsilon(t)$ and the normality of the distribution law, the sum of the errors of positive and negative products is balanced. However, as the latent period of the accident begins, when the noise $\varepsilon_2(i\Delta t)$ correlated with the useful signal emerges, and the sum noise is equal to

$$\varepsilon(i\Delta t) = \varepsilon_1(i\Delta t) + \varepsilon_2(i\Delta t),$$

these conditions are not fulfilled and the error of the sought-for estimates differs from zero. They can be calculated from the expressions

$$\begin{cases} \lambda_{a_n} = \sum_{i=1}^{N^+} \int_{t_1}^{t_{i+1}} \varepsilon(t)\cos n\omega t\,dt - \sum_{i=1}^{N^-} \int_{t_1}^{t_{i+1}} \varepsilon(t)\cos n\omega t\,dt \\ \lambda_{b_n} = \sum_{i=1}^{N^+} \int_{t_1}^{t_{i+1}} \varepsilon(t)\sin n\omega t\,dt - \sum_{i=1}^{N^-} \int_{t_1}^{t_{i+1}} \varepsilon(t)\sin n\omega t\,dt \end{cases}, \tag{4.3}$$

where t_1 and t_{i+1} are the start and the end of the i-th positive half-cycle of $\cos n\omega t$ and $\sin n\omega t$, t_{i+1} and t_{i+2} are the start and the end of the negative half-cycle of $\cos n\omega t$ and $\sin n\omega t$, N^+ and N^- are the number of the positive and negative half-cycles of $\cos n\omega t$ and $\sin n\omega t$, respectively.

In view of the above, let us first consider the problem of calculating the estimates a_n and b_n from Expressions (4.3). Assume that the observation time T of the realization of the noisy signal $g(t) = X(t) + \varepsilon(t)$ is chosen sufficiently large. Also assuming that the functions $X(t)$ and $\varepsilon(t)$ are sampled stationary centered random signals with zero-value mathematical expectations, the formulas for calculating the estimates of the coefficients a_n and b_n can be represented as

$$a_n = \frac{2}{N} \sum_{i=1}^{N} [X(i\Delta t) + \varepsilon(i\Delta t)]\cos n\omega(i\Delta t) = \frac{2}{N} \sum_{i=1}^{N} [X(i\Delta t)]\cos n\omega(i\Delta t)$$

$$+ \left[\frac{2}{N} \left[\sum_{i=1}^{N^+} \varepsilon(i\Delta t)\cos n\omega(i\Delta t) \right] - \sum_{i=1}^{N^-} \varepsilon(i\Delta t)\cos n\omega(i\Delta t) \right]$$

4.1 Algorithms and Technologies for Calculating the Errors ...

$$= \frac{2}{N} \sum_{i=1}^{N} \left[X(i\Delta t) \cos n\omega(i\Delta t) + \lambda_{a_n} \right], \tag{4.4}$$

$$b_n = \frac{2}{N} \sum_{i=1}^{N} [X + \varepsilon(i\Delta t)] \sin n\omega(i\Delta t) = \frac{2}{N} \sum_{i=1}^{N} [X(i\Delta t) \sin n\omega(i\Delta t)]$$

$$+ \left[\frac{2}{N} \left[\sum_{i=1}^{N^+} \varepsilon(i\Delta t) \sin n\omega(i\Delta t) \right] - \sum_{i=1}^{N^-} \varepsilon(i\Delta t) \sin n\omega(i\Delta t) \right]$$

$$= \frac{2}{N} \sum_{i=1}^{N} \left[X(i\Delta t) \sin n\omega(i\Delta t) + \lambda_{b_n} \right], \tag{4.5}$$

where $X(i\Delta t)$ and $\varepsilon(i\Delta t)$ are the samples of the signal $X(t)$ and the noise $\varepsilon(t)$, respectively, at the sampling instant $t_0, t_1, \ldots, t_i, \ldots, t_N$ with the interval Δt.

It is clear that in this case, when the above-mentioned conditions hold, the error of positive N^+ and negative N^- pair products of $\varepsilon(i\Delta t) \cos n\omega(i\Delta t)$ and $\varepsilon(i\Delta t) \sin n\omega(i\Delta t)$ will be balanced out. However, in those cases when the object goes into an emergency state, as a correlation appears between the useful signal $X(i\Delta t)$ and the noise $\varepsilon(i\Delta t)$, the errors λ_{a_n} and λ_{b_n} appear, the values of the errors increasing as the degree of correlation grows. As a result, in some cases, the estimates of the errors λ_{a_n} and λ_{b_n} caused by the effects of the noise $\varepsilon(i\Delta t)$ turn out to be commensurate with the sought-for coefficients a_n and b_n, which often leads to errors in the results of the monitoring of the onset of changes in the technical condition of control objects.

In view of the above, let us now consider one possible way of balancing the positive and negative errors of the estimates a_n and b_n of the respective paired products [1]. Assume that the samples of the noise $\varepsilon(i\Delta t)$ of the sum signal $g(i\Delta t)$ are known. In this case, the value of the error $\lambda(i\Delta t)$ of each pair product $g(i\Delta t) \cos n\omega(i\Delta t)$ and $g(i\Delta t) \sin n\omega(i\Delta t)$ can be calculated from the formulas

$$\lambda_{a_n}(i\Delta t) = |\varepsilon(i\Delta t) \cos n\omega(i\Delta t)|, \quad \lambda_{b_n}(i\Delta t) = |\varepsilon(i\Delta t) \sin n\omega(i\Delta t)|. \tag{4.6}$$

The estimate of the mean value of the error $\lambda(i\Delta t)$ can be calculated from the formula

$$\bar{\lambda}_{a_n}(i\Delta t) = \overline{|\varepsilon(i\Delta t) \cos n\omega(i\Delta t)|}, \quad \bar{\lambda}_{b_n}(i\Delta t) = \overline{|\varepsilon(i\Delta t) \sin n\omega(i\Delta t)|}, \tag{4.7}$$

that can be used to calculate the approximate values of the errors of the estimates a_n and b_n in the latent period of the object's emergency state from the expressions

$$\lambda_{a_n} = \left[\frac{N^+_{a_n} - N^-_{a_n}}{N} \bar{\lambda}_a(i\Delta t) \right], \quad \lambda_{b_n} = \left[\frac{N^+_{b_n} - N^-_{b_n}}{N} \bar{\lambda}_b(i\Delta t) \right], \tag{4.8}$$

where N^+ and N^- are number of positive and negative errors of pair products $g(i\Delta t)\cos n\omega(i\Delta t)$ and $g(i\Delta t)\sin n\omega(i\Delta t)$, respectively.

Thus, by balancing the positive and negative errors, it is possible to increase the reliability of the results of calculating the estimates a_n and b_n from the formulas

$$a_n^* = \frac{2}{N}\sum_{i=1}^{N} g(i\Delta t)\cos n\omega(i\Delta t) - \lambda_{a_n}, \qquad (4.9)$$

$$b_n^* = \frac{2}{N}\sum_{i=1}^{N} g(i\Delta t)\sin n\omega(i\Delta t) - \lambda_{b_n}, \qquad (4.10)$$

the use of which allows improving the reliability of the results of control during object's operation both in the normal mode and in the emergency mode.

However, to use Expressions (4.9) and (4.10), it is necessary to calculate the errors λ_{a_n} and λ_{b_n}, which, in turn, requires calculating the samples of the noise $\varepsilon(i\Delta t)$.

4.2 Algorithms for Calculating the Estimates of Spectral Characteristics of the Noise in the Latent Period of Object's Emergency State

As mentioned in the previous sections of this book, the initiation of faults and the dynamics of their development are accompanied by the emergence of the noise $\varepsilon_2(i\Delta t)$ correlated with the useful signal $X(i\Delta t)$. The noise $\varepsilon_2(i\Delta t)$ is added to the noise $\varepsilon_1(i\Delta t)$, forming the sum noise $\varepsilon(i\Delta t)$

$$\varepsilon(i\Delta t) = \varepsilon_1(i\Delta t) + \varepsilon_2(i\Delta t),$$

which in the latent period of accidents correlates with the useful signal.

Therefore, when solving the problem of controlling the beginning and development of this process, it is expedient to use estimates of the spectral characteristics of the sum noise $(i\Delta t)$ as informative attributes. An analysis of possible solutions to this problem has shown [2–5] that it is possible to replace unmeasurable samples of the noise $\varepsilon(i\Delta t)$ with their approximate equivalent values, and for this purpose it is possible and appropriate to use the technology for calculating the estimate of the noise variance D_ε from the expression

$$D_\varepsilon \approx \frac{1}{N}\sum_{i=1}^{N}[g(i\Delta t)g(i\Delta t) + g((i+2)\Delta t)g(i\Delta t)$$
$$- 2g(i\Delta t)g((i+1)\Delta t)], \qquad (4.11)$$

which can also be represented in the form

4.2 Algorithms for Calculating the Estimates of Spectral …

$$\frac{1}{N}\sum_{i=1}^{N}\varepsilon^2(i\Delta t) \approx \frac{1}{N}\sum_{i=1}^{N}g(i\Delta t)[g(i\Delta t)+g((i+2)\Delta t)-2g((i+1)\Delta t].\quad(4.12)$$

Thus, taking the notation

$$\varepsilon'(i\Delta t)=g(i\Delta t)[g(i\Delta t)+g((i+2)\Delta t)-2g((i+1)\Delta t]\quad(4.13)$$

$$\operatorname{sgn}\varepsilon'(i\Delta t)=\begin{cases}+1 & \text{when } \varepsilon'(i\Delta t)>0\\ 0 & \text{when } \varepsilon'(i\Delta t)=0\\ -1 & \text{when } \varepsilon'(i\Delta t)<0\end{cases}\quad(4.14)$$

the formula for calculating the equivalent values of the samples of the noise $\varepsilon(i\Delta t)$ can be represented as

$$\varepsilon(i\Delta t)\approx \varepsilon^e(i\Delta t)=\operatorname{sgn}\varepsilon'(i\Delta t)\sqrt{|g(i\Delta t)[g(i\Delta t)+g((i+2)\Delta t)-2g((i+1)\Delta t]|}$$
$$=\operatorname{sgn}\varepsilon'(i\Delta t)\sqrt{|\varepsilon'(i\Delta t)|}.\quad(4.15)$$

And, assuming that the expression

$$D_\varepsilon = \frac{1}{N}\sum_{i=1}^{N}\varepsilon^2(i\Delta t)\approx \frac{1}{N}\sum_{i=1}^{N}\varepsilon^{e2}(i\Delta t)$$
$$=\frac{1}{N}\sum_{i=1}^{N}|g(i\Delta t)[g(i\Delta t)+g((i+2)\Delta t)-2g((i+1)\Delta t]|,\quad(4.16)$$

is true, the formula for calculating the mean value $\bar{\varepsilon}(i\Delta t)$ of the samples of the noise $\varepsilon(i\Delta t)$ can be reduced to calculating the mean value of the equivalent samples of the noise $\varepsilon^e(i\Delta t)$, i.e.,

$$\bar{\varepsilon}(i\Delta t)\approx \overline{\varepsilon^e}(i\Delta t)=\frac{1}{N}\sum_{i=1}^{N}\varepsilon^e(i\Delta t).\quad(4.17)$$

Consequently, Expression (4.12) can be written as

$$\bar{\lambda}_{a_n}(i\Delta t)=\left|\overline{\varepsilon^e}(i\Delta t)\cos n\omega(i\Delta t)\right|,\bar{\lambda}_{b_n}(i\Delta t)=\left|\overline{\varepsilon^e}(i\Delta t)\sin n\omega(i\Delta t)\right|\quad(4.18)$$

To this end, taking into account Expressions (4.10)–(4.16), the algorithms of the spectral analysis of the noise $\varepsilon(i\Delta t)$ of the noisy signal $g(i\Delta t)$ can be written as

$$a_{n_\varepsilon}\approx \frac{2}{N}\sum_{i=1}^{N}\varepsilon^e(i\Delta t)\cos n\omega(i\Delta t),\quad(4.19)$$

$$b_{n_\varepsilon} \approx \frac{2}{N} \sum_{i=1}^{N} \varepsilon^e(i\Delta t) \sin n\omega(i\Delta t). \tag{4.20}$$

It is easy to see that, taking into account notation (4.10) and (4.12), Expressions (4.19) and (4.20), i.e., the formulas for calculating the estimates of the spectral characteristics of the noise, can be written in the form

$$a_{n_\varepsilon} \approx \frac{2}{N} \sum_{i=1}^{N} \operatorname{sgn} \varepsilon'(i\Delta t) \sqrt{|g(i\Delta t)[g(i\Delta t) + g((i+2)\Delta t) - 2g((i+1)\Delta t]|}$$

$$\times \cos n\omega(i\Delta t) = \frac{2}{N} \operatorname{sgn} \varepsilon'(i\Delta t) \sqrt{|\varepsilon'(i\Delta t)|} \cos n\omega(i\Delta t), \tag{4.21}$$

$$b_{n_\varepsilon} \approx \frac{2}{N} \sum_{i=1}^{N} \operatorname{sgn} \varepsilon'(i\Delta t) \sqrt{|g(i\Delta t)[g(i\Delta t) + g((i+2)\Delta t) - 2g((i+1)\Delta t]|}$$

$$\times \sin n\omega(i\Delta t) = \frac{2}{N} \sum_{i=1}^{N} \operatorname{sgn} \varepsilon'(i\Delta t) \sqrt{|\varepsilon'(i\Delta t)|} \sin \omega(i\Delta t). \tag{4.22}$$

Thus, the estimates a_n^* and b_n^* obtained from Expressions (4.9) and (4.10) increase the reliability of the results of spectral analysis, which ensures the reliability of the results of control in the object's normal operation mode. However, by using Expressions (4.21) and (4.22), it is possible to control the latent period of the emergency state. These algorithms provide an opportunity to register the beginning of the latent period of accidents, since at the beginning of an emergency state, the estimates a_{n_ε} and b_{n_ε} differ from zero.

It is common knowledge that despite the influence of various factors hampering the provision of accident-free operation of objects, modern control systems ensure their normal functioning. However, there is always a danger of object going into the latent period of an emergency state. Therefore, due to the extreme importance of ensuring the reliability and validity of the results of controlling the current state, it is expedient to duplicate traditional control algorithms with spectral algorithms for noise control of the beginning of initiation and development dynamics of faults. Consequently, we can assume that the use of the algorithms and technology for forming the informative attributes from the spectral estimates of the noise combined with traditional algorithms will improve the reliability and validity of ensuring object's operation in the normal mode. The increase in the value of estimates in time is taken as a result of the dynamics of the development of accidents, since this can be caused by both $\varepsilon_2(i\Delta t)$ and $\varepsilon_1(i\Delta t)$.

Studies have shown that only the value of the estimate of the cross-correlation function between $X(i\Delta t)$ and $\varepsilon(i\Delta t)$ clearly depends on the effect of the noise $\varepsilon_2(i\Delta t)$. However, it is impossible to calculate the samples of the noise $\varepsilon_2(i\Delta t)$. According to the previous studies, by using the expression for calculating $R_{x\varepsilon}(\mu)$, we can form the analog $\varepsilon_2^a(i\Delta t)$ of the noise $\varepsilon_2(i\Delta t)$ from the formula

4.2 Algorithms for Calculating the Estimates of Spectral ...

$$R_{x\varepsilon}(\mu) = \frac{1}{N} \sum_{i=1}^{N} g(i\Delta t)[g(i+1)\Delta t + g(i+3)\Delta t - 2g(i+2)\Delta t]$$

$$= \frac{1}{N} \sum_{i=1}^{N} \varepsilon_2^a((i+1)\Delta t).$$

The formation of $\varepsilon_2^a(\mu = 1\Delta t)$ can be written as

$$\varepsilon_2^a(i\Delta t) = g(i\Delta t)[g(i+1)\Delta t + g(i+3)\Delta t - 2g(i+2)\Delta t].$$

When $\mu = 2\Delta t$, this expression will have the following form:

$$g(i\Delta t)[g(i+2)\Delta t + 2g(i+4)\Delta t - 2g(i+3)\Delta t] = \varepsilon_2^a((i+2)\Delta t).$$

When $\mu = m\Delta t$, this expression can be written in the generalized form

$$g(i\Delta t)[g(i+m)\Delta t + 2g(i+m+2)\Delta t - 2g(i+m+1)\Delta t] = \varepsilon_2^a(i+m)\Delta t.$$

Research shows that using the results of the spectral analysis of the analog $\varepsilon_2^a(i\Delta t)$ of the noise $\varepsilon(i\Delta t)$ at $\mu = 1, 2, 3, \ldots$, i.e.

$$\varepsilon_2^a(i+1)\Delta t; \varepsilon_2^a(i+2)\Delta t; \varepsilon_2^a(i+3)\Delta t, \ldots, \varepsilon_2^a(i+m)\Delta t,$$

we can control the development dynamics of an accident by means of the expressions

$$a^*_{1\varepsilon_2} \approx \frac{2}{N} \sum_{i=1}^{N} \varepsilon_2^a(i+1)\Delta t \cos n\omega(i\Delta t)],$$

$$b^*_{1\varepsilon_2} \approx \frac{2}{N} \sum_{i=1}^{N} \varepsilon_2^a(i+1)\Delta t \sin n\omega(i\Delta t),$$

$$a^*_{2\varepsilon_2} \approx \frac{2}{N} \sum_{i=1}^{N} \varepsilon_2^a(i+2)\Delta t \cos n\omega(i\Delta t),$$

$$b^*_{2\varepsilon_2} \approx \frac{2}{N} \sum_{i=1}^{N} \varepsilon_2^a(i+2)\Delta t \sin n\omega(i\Delta t),$$

$$\ldots\ldots\ldots\ldots\ldots\ldots\ldots\ldots\ldots\ldots\ldots\ldots\ldots$$

$$a^*_{n\varepsilon_2} \approx \frac{2}{N} \sum_{i=1}^{N} \varepsilon_2^a(i+m)\Delta t \cos n\omega(i\Delta t),$$

$$b^*_{n\varepsilon_2} \approx \frac{2}{N} \sum_{i=1}^{N} \varepsilon_2^a(i+m)\Delta t \sin n\omega(i\Delta t).$$

At the same time, under the influence of development dynamics of accidents, these estimates will first be different from zero, then, if the degree of accident develops, increase continually.

4.3 Spectral Technology of Noise Signaling for the Beginning of the Latent Period of Accidents

An analysis of spectral noise control technologies has demonstrated that during object's operation, the signaling of the beginning of the latent period of accidents is very important. For this purpose, as informative attributes, in addition to the above, it is also expedient to use the estimates $a_{n\varepsilon}^{**}$ and $b_{n\varepsilon}^{**}$ relay spectral characteristics of the noise $\varepsilon(i\Delta t)$ of the noisy signals $g(i\Delta t)$ which are calculated from the expressions

$$a_{n\varepsilon}^{**} = \frac{1}{N}\sum_{i=1}^{N} \operatorname{sgn} \varepsilon(i\Delta t) \cos n\omega(i\Delta t)$$

$$= \frac{2}{N}\sum_{i=1}^{N} \operatorname{sgn} \varepsilon'(i\Delta t)\sqrt{|\varepsilon'(i\Delta t)|} \cos n\omega(i\Delta t), \qquad (4.22)$$

$$b_{n\varepsilon}^{**} = \frac{1}{N}\sum_{i=1}^{N} \operatorname{sgn} \varepsilon(i\Delta t) \sin n\omega(i\Delta t)$$

$$= \frac{2}{N}\sum_{i=1}^{N} \operatorname{sgn} \varepsilon'(i\Delta t)\sqrt{|\varepsilon'(i\Delta t)|} \sin n\omega(i\Delta t). \qquad (4.23)$$

These studies have also demonstrated that for signaling the beginning of fault initiation, the technology of sign spectral noise analysis can also be used with the help of the expressions

$$a_{n\varepsilon}' = \frac{1}{N}\sum_{i=1}^{N} \operatorname{sgn} \varepsilon(i\Delta t)\operatorname{sgn} \cos n\omega(i\Delta t)$$

$$= \frac{2}{N}\sum_{i=1}^{N} \operatorname{sgn} \varepsilon'(i\Delta t)\sqrt{|\varepsilon'(i\Delta t)|}\operatorname{sgn} \cos n\omega(i\Delta t),$$

$$b_{n\varepsilon}' = \frac{1}{N}\sum_{i=1}^{N} \operatorname{sgn} \varepsilon(i\Delta t)\operatorname{sgn} \sin n\omega(i\Delta t)$$

$$= \frac{1}{N}\sum_{i=1}^{N} \operatorname{sgn} \varepsilon'(i\Delta t)\sqrt{|\varepsilon'(i\Delta t)|}\operatorname{sgn} \sin n\omega(i\Delta t). \qquad (4.24)$$

4.3 Spectral Technology of Noise ...

The expediency of using the technology of relay and sign spectral analysis for signaling the beginning of the latent period of accidents is due to the fact that they are easily implemented in terms of hardware. However, they do not allow registering the development dynamics of accidents, as the values of the samples of the equivalent interference $\varepsilon^e(i\Delta t)$ are affected both by the noise $\varepsilon_1(i\Delta t)$ and the noise $\varepsilon_2(i\Delta t)$.

Therefore, for signaling the development dynamics of accidents, it is advisable to use the technology for calculating the relay spectral analog $\varepsilon_2^a(i\Delta t)$ of the noise $\varepsilon_2(i\Delta t)$ from the expressions

$$a_{1\varepsilon_2}^* = \frac{2}{N} \sum_{i=1}^{N} \operatorname{sgn} \varepsilon_2^a(i\Delta t) \cos n\omega(i\Delta t),$$

$$b_{1\varepsilon_2}^* = \frac{2}{N} \sum_{i=1}^{N} \operatorname{sgn} \varepsilon_2^a(i\Delta t) \sin n\omega(i\Delta t),$$

$$a_{2\varepsilon_2}^* = \frac{2}{N} \sum_{i=1}^{N} \operatorname{sgn} \varepsilon_2^a((i+1)\Delta t) \cos n\omega(i\Delta t),$$

$$b_{2\varepsilon_2}^* = \frac{2}{N} \sum_{i=1}^{N} \operatorname{sgn} \varepsilon_2^a((i+1)\Delta t) \sin n\omega(i\Delta t),$$

$$\cdots\cdots\cdots\cdots\cdots\cdots\cdots\cdots\cdots\cdots\cdots\cdots\cdots\cdots\cdots\cdots$$

$$a_{n\varepsilon_2}^* = \frac{2}{N} \sum_{i=1}^{N} \operatorname{sgn} \varepsilon_2^a((i+m)\Delta t) \cos n\omega(i\Delta t),$$

$$b_{n\varepsilon_2}^* = \frac{2}{N} \sum_{i=1}^{N} \operatorname{sgn} \varepsilon_2^a((i+m)\Delta t) \sin n\omega(i\Delta t).$$

It is obvious that the signaling about the development dynamics of accidents can also be realized using the estimates of the sign spectral characteristics of the analog $\varepsilon_2^a(i\Delta t)$ of the noise $\varepsilon_2(i\Delta t)$ from the formulas

$$a_{1\varepsilon_2}^{**} = \frac{2}{N} \sum_{i=1}^{N} \operatorname{sgn} \varepsilon_2^a(i\Delta t) \operatorname{sgn} \cos n\omega(i\Delta t),$$

$$b_{1\varepsilon_2}^{**} = \frac{2}{N} \sum_{i=1}^{N} \operatorname{sgn} \varepsilon_2^a(i\Delta t) \operatorname{sgn} \sin n\omega(i\Delta t),$$

$$a_{2\varepsilon_2}^{**} = \frac{2}{N} \sum_{i=1}^{N} \operatorname{sgn} \varepsilon_2^a((i+1)\Delta t) \operatorname{sgn} \cos n\omega(i\Delta t),$$

$$b_{2\varepsilon_2}^{**} = \frac{2}{N} \sum_{i=1}^{N} \operatorname{sgn} \varepsilon_2^a((i+1)\Delta t) \operatorname{sgn} \sin n\omega(i\Delta t),$$

$$a^{**}_{n\varepsilon_m} = \frac{2}{N} \sum_{i=1}^{N} \operatorname{sgn} \varepsilon_2^a((i+m)\Delta t) \operatorname{sgn} \cos n\omega(i\Delta t),$$

$$b^{**}_{n\varepsilon_m} = \frac{2}{N} \sum_{i=1}^{N} \operatorname{sgn} \varepsilon_2^a((i+m)\Delta t) \operatorname{sgn} \sin n\omega(i\Delta t),$$

The use of these technologies makes it possible to signal the presence of development dynamics of accidents at $\mu = 1\Delta t$, $\mu = 2\Delta t$, $\mu = 3\Delta t$, ..., $\mu = m\Delta t$. At the same time, the reliability of the signaling is ensured by the duplication with analogous estimates of the relay correlation functions.

Therefore, due to the extreme importance of ensuring object's accident-free operation, it is advisable to control the beginning and dynamics of development of faults by duplicating with several technologies of the spectral noise control and noise signaling proposed above.

This will increase the reliability of object's operation in the normal mode.

4.4 Position-Binary Technology for the Control of the Beginning of the Latent Period of Accidents of Objects of Periodical Operation

It is known that experimental analysis of cyclic (periodic) processes in most cases employs spectral methods [6–10]. For instance, objects with equipment of reciprocating motion, objects with rotating nodes, some biological processes, etc., are cyclic. The importance of this problem is due to the fact that in real life such objects as cars, compressor stations, power stations, technical facilities working with electric motors, etc., are very widespread. Signals received from all these objects, as a rule, have a complex intermittent abruptly changing form and are accompanied by significant noises. At present, spectral methods and algorithms are widely used for their experimental investigation [10–17]. However, in some cases, they are not effective enough for these objects [1]. This is because, in order to adequately describe intermittent and abruptly changing periodic signals, it is often necessary to use a large number of harmonic components with corresponding amplitudes and frequencies. This significantly complicates the analysis and the use of the results obtained for decision-making [2–10]. Therefore, when solving problems of the control of the beginning of the latent period of accidents for the considered class of objects, there is a need to develop methods and algorithms that would allow simultaneous reduction of the number of components of the "spectrum" and increase in the reliability of the obtained results in comparison with the spectral method.

Let us consider in more detail the difficulties of control in the latent period of an emergency state for objects in the cyclic operation mode.

4.4 Position-Binary Technology for the Control of the Beginning ...

It is known that when spectral method algorithms are used for the analysis of the periodic signals $X(t)$ with bounded spectrum, they are resolved into harmonic components, using the expression

$$X(t) = \frac{a_n}{2} + \sum_{n=1}^{\infty} (a_n \cos n\omega t + b_n \sin n\omega t). \tag{4.25}$$

In Expression (4.1), a_n, b_n are the amplitudes of the cosine curve and the sine curve with the frequency $n\omega$, which are taken as informative attributes in controlling the beginning of accident initiation. It is known that the following inequality should hold true to provide accuracy of reconstruction of the signal $X(t)$.

In Expression (4.25) the coefficients a_n and b_n are the amplitudes of the cosine curve and the sine curve with the frequency $n\omega$, which are taken as informative attributes in controlling the beginning of accident initiation. The following inequality should hold true to provide accuracy of reconstruction of the signal $X(t)$:

$$\sum_{i=1}^{n} \lambda_i^2 \leq S, \tag{4.26}$$

where λ_i^2 are the squared deviations between the sum of the right-hand side of equality (4.25) and the samples of the signal $X(t)$ at the sampling moments $t_0, t_1, \ldots, t_i, \ldots, t_n$ with the interval Δt; S is the allowable value of the mean square deviation.

For intermittent and abruptly changing periodic signals, ensuring inequality (4.26) leads to an increase of the number of harmonic components, which therefore complicates the analysis of experimental data. In addition, for the case where the measurement information is the mixture of the useful signal $X(t)$ and the noise $\varepsilon(t)$, condition (4.26) holding true depends to a certain extent on the value of the spectrum of the noise $\varepsilon(t)$. In the existing methods in equality (4.25) the influence of noise is neglected and the error of the noise $\varepsilon(t)$ is equated to zero. However, for many cyclic processes, the influence of the noise on the accuracy of reconstruction of the source signal $X(t)$ turns out to be significant and should be taken into.

If we take into account that in the latent period of initiation and development of accidents, the spectrum of the noise $\varepsilon(t)$ changes continuously, then the difficulty of using the technology of traditional spectral analysis for solving the problem of control of object's technical condition in this period will become obvious. Therefore, it is necessary to create new spectral technologies that take into account the specific characteristics of signals received from periodic objects. Studies [6–10] have demonstrated that when solving this problem, the use of position-binary technology (PBT) in the analysis of cyclic noisy signals is expedient, so that reliable and valid results can be obtained both in the normal mode and in the latent period of an emergency state of periodic objects. In this regard, we will consider the possibilities of this technology in more detail.

In reality, in measuring of the signals $g(t)$, there is a minimum value of increment that can be provided for by the device in use and depends on its resolving capacity. Denote that minimum value of increment by Δx. Therefore, during measuring of the signal, the number of its discrete values will be equal to

$$m = \frac{x}{\Delta x} + 1. \tag{4.27}$$

In the process of analog-to-digital conversion of the periodic signal $g(t)$, its amplitude sampling occurs at each sampling interval Δt, i.e., the range of its possible variations is divided into $m_{\max}$ of quantization steps, and the value of the signal that gets into the m-th interval, with

$$m\Delta X - \frac{\Delta X}{2} \leq X(t) \leq m\Delta x + \frac{\Delta x}{2}, \tag{4.28}$$

belongs to the center of the interval $m\Delta X$. In this case, values of binary codes of corresponding bits q_k of the samples of the signal $X(i\Delta t)$ with the sampling interval Δt are calculated based on the following algorithms [6–10]:

$$q_k(i\Delta t) = \begin{cases} 1 & \text{when } x_{\text{rem}(k)}(i\Delta t) \geq \Delta x 2^k; \\ 0 & \text{when } x_{\text{rem}(k)}(i\Delta t) < \Delta x 2^k; \end{cases} \tag{4.29}$$

$$x_{\text{rem}(k)}(i\Delta t) = x_k(i\Delta t) - \left[q_{k+1}(i\Delta t) + q_{k+2}(i\Delta t) + \cdots + q_{(n-1)}(i\Delta t)\right],$$

where

$$X(i\Delta t) > 2^n; x_{\text{rem}(n-1)}(i\Delta t) = X(i\Delta t), n \geq \log \frac{x_{\max}}{\Delta x},$$

$$k = n - 1, n - 2, \ldots, 1, 0.$$

In accordance with this algorithms, the equality $x_{\text{rem}(n-1)}(i\Delta t) = X(i\Delta t)$ is assumed at each sampling interval Δt, and according to condition (4.29), the signals $q_k(i\Delta t)$ form in the form of code 1 or 0 iteratively. At the first step, $X(i\Delta t)$ is compared with the value $2^{n-1}\Delta x$. According to (4.29), if $X(i\Delta t) \geq 2^{n-1}\Delta x$, then the bit $q_{n-1}(i\Delta t)$ is equated to unit and the value of the remainder $x_{\text{rem}(n-2)}$ is calculated from the difference

$$X(i\Delta t) - 2^{n-1}\Delta x = x_{\text{rem}(n-2)}.$$

In the case when $X(i\Delta t) < 2^{n-1}\Delta x$, the bit $q_{n-1}(i\Delta t)$ is equated to zero, and the difference remains unchanged. The same thing occurs in the next iteration. As a result, in every conversion cycle with the sampling interval Δt, the signal $X(i\Delta t)$ is as if decomposed into the signals $q_k(i\Delta t)$, which assume values 1 or 0 and have weights corresponding to their positions. The codes remain unchanged as long as the value of the source signal $g(i\Delta t)$ does not change in the sampling process. Let us

4.4 Position-Binary Technology for the Control of the Beginning ...

from now onwards call these signals position-binary pulse signals (PBPS). Position-binary technology (PBT) of analysis of the noisy signal is the aggregate of successive processing procedures based on decomposition of the continuous signal into PBPS.

According to algorithm (4.29), the width of PBPS will be proportional to the number of Δt, when $q_k(i\Delta t)$ remains unchanged. Depending on the shape of $X(i\Delta t)$, the same signal $q_k(i\Delta t)$ can change its value several times at corresponding time spans. Note that $T_{k1_1}, T_{k1_2}, \ldots$ here correspond to the time spans when the condition $q_k(i\Delta t) = 2^k(\Delta x = 1)$ is fulfilled; $T_{k0_1}, T_{k0_2}, \ldots$ correspond to the time spans when the condition $q_k(i\Delta t) = 2^k(\Delta x = 0)$ is fulfilled. Naturally, if the object's technical condition operating in the cyclic mode remains unchanged, then combinations of the time spans $T_{k1_1}, T_{k0_1}, T_{k1_2}, T_{k0_2}, \ldots$ of PBPS in each cycle will be constant values and reiterate. Otherwise, they change as well. According to Expression (4.29), the sum of all PBPS in each cycle will be equal to the source signal, i.e.,

$$g(i\Delta t) \approx q_{n-1}(i\Delta t) + q_{n-2}(i\Delta t) + \ldots + q_1(i\Delta t) + q_0(i\Delta t) = g^*(i\Delta t). \quad (4.30)$$

Assume that the cycle time of the control object and, therefore, the cycle time of the analyzed signal, is 15 µs, the sampling interval is 1 µs, i.e., $T_c = 15$ µs, $\Delta t = 1$ µs. Assume that PBPS $q_3(i\Delta t)$ in one cycle assumes the following states: 000111100110000. In the first case, the parameters of the signal $q_3(i\Delta t)$ will be represented in the following form: $T_3 = 3.0; T_3 = 4.1; T_3 = 2.0; T_3 = 2.1; T_3 = 4.0$. This means that the width of the unit and zero states of the signal $q_3(i\Delta t)$ during the cycle has the time spans 3 µs–0.4 µs–1.2 µs–0.2 µs–1.4 µs–0, respectively.

Each $q_k(i\Delta t)$ can be regarded as a separate signal, and the combinations of sequences of the time spans when $q_k(i\Delta t)$ are in the state of unit or zero, can be regarded as pulse-width signals. Due to this, those PBPS $q_k(i\Delta t)$ will be periodic rectangular pulses with the corresponding unit T_1 and zero T_0 half-periods.

And at the moments t_i, the difference between the true value of the source signal $g(i\Delta t)$ and the sum of PBPS will be equal to

$$g(i\Delta t) - g^*(i\Delta t) = \lambda(i\Delta t). \quad (4.31)$$

Taking into account Expression (4.28), the following inequality can be written:

$$\lambda(i\Delta t) \leq \pm \frac{\Delta x}{2}.$$

If we assume that when the signals $q_k(i\Delta t)$ form, the value of the error $\lambda(i\Delta t)$ complies with the equiprobable distribution law, then the following equality can be regarded as true

$$P\left[\lambda_i < \frac{\Delta x}{2}\right] \approx P\left[\lambda_i > \frac{\Delta x}{2}\right], \quad (4.32)$$

where P is the probability sign.

Therefore, according to (4.31) and (4.32), the sum of squared deviations λ_i at the moments $t_0, t_1, \ldots, t_i, \ldots$ with allowance for their sign will be close to zero. Thus, inequality (4.26) can be written as follows:

$$\sum_{i=1}^{n} \lambda^2(i\Delta t) \leq \Delta x.$$

In accordance with this inequality, when the signal $g(i\Delta t)$ is represented as the sum of PBPS, the mean square deviation will not exceed the value Δx. Due to this, a change in object's technical condition will lead to a change in the corresponding samples of the signal $g(i\Delta t)$ by a value exceeding Δx, and it will reflect on its corresponding bits $q_k(i\Delta t)$. Therefore, already at the initial stage of the change in the process of PBPS formation in the form of the combination of the corresponding time spans $q_{n-1}(i\Delta t), q_{n-2}(i\Delta t), \ldots, q_0(i\Delta t)$ of the current cycle, the difference from their analogous parameters in previous cycles will be revealed, which will allow one to form and present the information on the change in control object's state. This only requires calculating the mean frequency $\langle f_k \rangle$ and the period $\langle T_k \rangle$ for each PBSP. The algorithms for their calculation are easily implemented in practice, since each position-random function assumes only two values. It is intuitively clear that for random and periodic noisy signals $g(i\Delta t)$ the estimate of the mean value of zero and unit half-periods of the position signals $q_k(i\Delta t)$, given the sufficient observation time T, can be calculated from the formula

$$\langle T_{q_k} \rangle = \langle T_{1q_k} \rangle + \langle T_{0q_k} \rangle, \tag{4.33}$$

where

$$\langle T_{1q_k} \rangle = \frac{1}{\gamma} \sum_{j=1}^{\gamma} T_{1q_{kj}}, \quad \langle T_{0q_k} \rangle = \frac{1}{\gamma} \sum_{j=1}^{\gamma} T_{0q_{kj}}. \tag{4.34}$$

Here γ is the number of unit and zero half-periods of PBPS in the observation time T, j is the is ordinal number of the q_k-the position of PBPS.

It is demonstrated in [1] that for the sufficient observation time, the estimate of the duration of the periods $\langle T_k \rangle$ and the mean frequency f_{q_k} of PBPS will be nonrandom values. Therefore, they can be used to control the beginning of changes in object's technical condition. And due to the simplicity of their calculation, they can significantly simplify solving of control problems, which are traditionally solved by means of estimates of correlation or spectral characteristics of random processes. For instance, in cyclic object's normal stable technical condition, sets of combinations of mean frequencies of PBPS $q_k(i\Delta t)$ will form from the signal $g(t)$. Obviously, a change in object's technical condition will lead to changes in the combinations of the estimates of their mean frequencies $\bar{f}_{q_0}, \bar{f}_{q_1}, \ldots, \bar{f}_{q_m}$, which are calculated from the expressions

4.4 Position-Binary Technology for the Control of the Beginning ... 73

$$\bar{f}_{q_0} = \frac{1}{\langle T_{q_0}\rangle}, \bar{f}_{q_1} = \frac{1}{\langle T_{q_1}\rangle}, \bar{f}_{q_2} = \frac{1}{\langle T_{q_2}\rangle}, \ldots, \bar{f}_{q_m} = \frac{1}{\langle T_{q_m}\rangle}. \quad (4.35)$$

The relation between the estimates T_{1q_k} and T_{0q_k} is also a nonrandom value

$$k_{q_0} = \frac{\langle T_{1q_0}\rangle}{\langle T_{0q_0}\rangle}, k_1 = \frac{\langle T_{1q_1}\rangle}{\langle T_{0q_1}\rangle}, k_2 = \frac{\langle T_{1q_2}\rangle}{\langle T_{0q_2}\rangle}, \ldots, k_m = \frac{\langle T_{1q_m}\rangle}{\langle T_{0q_m}\rangle}. \quad (4.36)$$

Therefore, forming sets of informative attributes from the combinations of frequencies of PBPS $\bar{f}_{q_0}, \bar{f}_{q_1}, \bar{f}_{q_2}, \ldots, \bar{f}_{q_m}$ and the relations $k_{q_0}, k_{q_1}, k_{q_2}, \ldots, k_m$, we can solve the problem of control of changes in object's technical condition.

Let us consider the possibility of using the relationship between the beginning of the latent period of accidents and the appearance of the noise $\varepsilon_2(i\Delta t)$ with the application of the position-binary technology [6–10]. It is known that an abruptly changing high-frequency noise appears in many cyclic objects at the beginning of defect initiation, for instance, on cranes, conveyors, agricultural equipment, etc. In some cases, the beginning of the latent period of a heart disease also manifests itself as an abruptly changing noise in the electrocardiogram. In view of the above, let us consider this issue in more detail. As was indicated earlier, in encoding of continuous signals, the values of binary codes of the corresponding bits $q_k(i\Delta t)$ of samples of the signal $g(i\Delta t)$ in each sampling interval Δt the signals $q_k(i\Delta t)$ form as codes 1 or 0 iteratively. At the first step, the samples of $X(i\Delta t)$ are compared with the value $2^{n-1}\Delta g$. At $g(i\Delta t) \geq 2^{n-1}\Delta g$, the value of $q_{n-1}(i\Delta t)$ is equated to the unit, and the value of the remainder $g_{\text{rem}(n-2)}(i\Delta t)$ is calculated from the difference

$$g(i\Delta t) - 2^{n-1}\Delta g = g_{\text{rem}(n-2)}(i\Delta t) \quad (4.37)$$

The sequence of these signals $q_k(i\Delta t)$ is PBPS, the sum of which will be equal to the source signal.

In this case, in the latent period of accidents, in the process of analog-to-digital conversion, the noise $\varepsilon_2(i\Delta t)$ forms as short-time pulses $q_{\varepsilon k}(i\Delta t)$, the duration of which is many times shorter than the length of the higher position signals $q_k(i\Delta t)$.

It is demonstrated in [18] that for the sufficient observation time T the ratio of the number $N_{\varepsilon k}$ of the signals $q_{\varepsilon k}(i\Delta t)$ to the total number N_{q_0} of the position-pulse signals $q_0(i\Delta t)$

$$K_{q_{\varepsilon 0}} = \frac{N_{\varepsilon_0}}{N_{q_0 k}} \quad (4.38)$$

is a nonrandom value. Our studies have showed that in object's normal mode of operation, if its state is stable, the coefficient $K_{q_{\varepsilon 0}}$ remains stable. At the same time, the number of the position signals N_{ε_0} increases during the time T from the starting moment of object's transition to an emergency state. Starting from this time, the values of the coefficients K_{q_0} also change abruptly. Consequently, they are informative attributes and when solving problems of controlling the beginning of the latent period of accidents they can be used in combination with other informative attributes to increase the reliability of the results of control of object's technical condition in the latent period of an emergency state.

Our studies have also shown that for many technical facilities, where the use of traditional methods for controlling the beginning of accidents does not yield satisfactory results, the use of the relationship between the beginning of accidents and the change in the coefficients $K_{q_{\varepsilon 0}}$ and other noise characteristics obtained by the position-binary technology gives reliable results. For instance, an analysis of signals obtained during drilling, operation of compressor stations, etc., shows that such simple noise characteristics as $K_{q_{\varepsilon 0}}$ contain a significant amount of useful information, which allows to register the beginning of the process of accident initiation.

4.5 Position-Selective Technology for Calculating the Sampling Interval of the Noise in the Latent Period of Object's Emergency State

It is known that calculating the sampling interval of analog signals in practice is associated with considerable difficulties.

The problem becomes even more complicated if the noise of a signal needs to be analyzed. It is shown in [1, 6–10] that it is possible to use the position-binary technology for this purpose. Let us first consider the possibility of calculating the sampling interval Δt_ε of the source noisy signal $g(i\Delta t)$ by means of the frequency properties of PBPS. Assume that the analyzed signal is subjected to analog-to-digital conversion with the frequency f_t with knowingly small quantization step in time Δt_t, so that the following equality holds:

$$P[[X(i\Delta t)] \approx [X((i+1)\Delta t)]] \approx 1.$$

The values of binary codes of the samples of $g(i\Delta t)$ at each successive quantization step $g((i+1)\Delta t)$ will essentially repeat themselves. Thus, the frequency f_{q_0} of low-order PBPS $q_0(t)$, which can be calculated from the formula

$$f_{q_0} = \frac{1}{\langle T_{q_0} \rangle},$$

will be significantly less than the current sampling frequency f_t. Therefore, the following inequality will take place between the frequency f_T found by the traditional method and the current frequency f_t:

$$f_t \gg f_T \qquad (4.39)$$

4.5 Position-Selective Technology for Calculating the Sampling ...

It is intuitively clear that the estimate of the mean value of $\bar{f}_{q_0}$ for all realizations of the same stationary random signal or cyclic signal will be a stable value.

Hence, the estimate of the mean value of $\bar{f}_{q_0}$ can be calculated for the analyzed signal by selecting the value of f_t so that condition (4.39) holds true.

Here, in the process of the analog-to-digital conversion of each PBPS, condition (4.39) for calculating the sought-for frequency can be represented as follows:

$$f_t \geq f_{q_0}.$$

Following from this condition, the sampling interval Δt_ε for the noise of the noisy signal Δt_ε can be selected in accordance with the inequality

$$\Delta t_\varepsilon \leq \frac{1}{f_{q_0}}.$$

In that case, to calculate f_{q_0}, it is necessary to calculate the mean period of pulses of the low-order PBPS $\langle T_{q_0} \rangle$ and their mean repetition frequency, using samples of the analyzed signal after its conversion and recording it in the memory with the frequency f_t, from the expressions

$$\langle T_{q_0} \rangle = \langle T_{1q_0} \rangle + \langle T_{0q_0} \rangle,$$

$$f_{q_0} = \frac{1}{\langle T_{q_0} \rangle}, \qquad (4.40)$$

where $\langle T_{1q_0} \rangle$ and $\langle T_{0q_0} \rangle$ are calculated from the expressions

$$\langle T_{1q_0} \rangle = \frac{1}{\gamma} \sum_{j=1}^{\gamma} T_{1q_0 j} \quad \text{and} \quad \langle T_{0q_0} \rangle = \frac{1}{\gamma} \sum_{j=1}^{\gamma} T_{0q_0}. \qquad (4.41)$$

Our experimental studies show the estimates of the mean frequency of low-order bits of PBPS, in reality, are correlated with the spectrum of the noise $\varepsilon(i\Delta t)$. Therefore, the sampling interval Δt of the useful signal $X(t)$ is to be calculated from the frequency characteristics of high-order PBPS $q_2(i\Delta t)$ or $q_3(i\Delta t)$, i.e., from mean values of duration of their unit T_{k1} and zero T_{k0} of half-periods. They are also easy to calculate by averaging the time spans in accordance with Formulas (4.40), (4.41), i.e.,

$$\langle T_k \rangle = \langle T_{1q_3} \rangle + \langle T_{0q_3} \rangle, \langle T_{1q_3} \rangle = \frac{1}{\gamma} \sum_{j=1}^{\gamma} T_{1q_3 j}, \langle T_{0q_3} \rangle = \frac{1}{\gamma} \sum_{j=1}^{\gamma} T_{1q_3 j}. \qquad (4.42)$$

Our experiments have shown that the following approximate equality takes place for the useful signals $X(i\Delta t)$ with the normal law of distribution between the mean values of durations of periods of PBPS $q_0(i\Delta t), q_1(i\Delta t), q_2(i\Delta t), q_3(i\Delta t)$

$$\langle T_{q_0}\rangle \approx \frac{1}{2}\langle T_{q_1}\rangle, \langle T_{q_1}\rangle = \frac{1}{2}\langle T_{q_2}\rangle, \langle T_{q_2}\rangle = \frac{1}{2}\langle T_{q_3}\rangle.$$

Therefore, the sampling step Δt_x can also be calculated by means of the mean period of pulses of high-order bits of PBPS. And to calculate Δt_t, the formula

$$\Delta t_t \leq \frac{1}{2^k f_k},$$

can be used instead of the expression

$$\Delta t_t \geq \frac{1}{f_{q_0}},$$

where $k = 1, 2, 3$.

For instance, the formula for calculating the sampling interval Δt_t by means of the q_3-th PBPS can be written as

$$\Delta t_t \leq \frac{1}{2 f_3}. \tag{4.43}$$

It is obvious from the above that by using Expressions (4.40)–(4.43), the sampling interval Δt_t can be calculated in the process of analog-to-digital conversion. In this case, the software calculation of the sampling interval Δt_ε and Δt_x, according to the above algorithms, comes down to the following:

1. During the observation time T with the excess frequency f_t, the source signal $g(i \Delta t)$ is converted into a digital signal via analog-to-digital conversion and the file of its samples forms;
2. $\langle T_{q_0}\rangle$ and $\langle T_{q_3}\rangle$ are calculated from Formula (4.42)

$$\langle T_{q_0}\rangle = \langle T_{1q_0}\rangle + \langle T_{0q_0}\rangle,$$
$$\langle T_{q_3}\rangle = \langle T_{1q_3}\rangle + \langle T_{0q_3}\rangle;$$

3. f_{q_0} and f_{q_3} are calculated from Formula (4.40)

$$f_{q_0} = \frac{1}{\langle T_{q_0}\rangle},$$

$$f_{q_3} = \frac{1}{\langle T_{q_3}\rangle};$$

4. Δt_ε and Δt_x are calculated from the formulas

$$\Delta t_\varepsilon \leq \frac{1}{2^0 f_{q_0}},$$

$$\Delta t_x \leq \frac{1}{2^2 f_{q_3}}.$$

In solving the problem of control of the beginning and development dynamics of accidents, due to the calculation of Δt_ε, it is possible to analyze the noises $\varepsilon(i\Delta t)$ of the noisy signals $g(i\Delta t)$ as a carrier of diagnostic information.

References

1. Aliev T (2007) Digital noise monitoring of defect origin. Springer, Boston. https://doi.org/10.1007/978-0-387-71754-8
2. Aliev TA, Ali-Zade TA (1999) Robust algorithms for spectral analysis of the technological parameters of industrial plants. Autom Control Comput Sci 33(5):38–44
3. Aliev TA, Guliev QA, Rzaev AH et al (2009) Position-binary and spectral indicators of microchanges in the technical states of control objects. Autom Control Comput Sci 43(3):156–165. https://doi.org/10.3103/S0146411609030067
4. Aliev TA (2001) A robust technology for improving correlation and spectral characteristic estimators, correlation matrix conditioning, and identification adequacy. Autom Control Comput Sci 35(4):10–19
5. Aliev TA, Alizadeh TA (2000) Robust technology for calculation of the coefficients of the Fourier series of random signals. Autom Control Comput Sci 34(4):18–26
6. Aliev TA, Nusratov OK (1998) Algorithms for the analysis of cyclic signals. Autom Control Comput Sci 32(2):59–64
7. Aliev TA, Nusratov OK (1998) Position-width-impulse analysis of cyclic and random signals. Optoelectron Instrum Data Process 2:85–91
8. Aliev TA, Nusratov OK (1998) Position-width-pulse analysis and sampling of random signals. Autom Control Comput Sci 32(5):44–48
9. Aliev TA, Nusratov OK (1998) Pulse-width and position method of diagnostics of cyclic processes. J Comput Syst Sci Int 37(1):126–131
10. Aliev TA, Mamedova UM (2003) Positional binary methodology for extraction of interference from noisy signals. Autom Control Comput Sci 37(2):12–19
11. Bendat JS, Piersol AG (2010) Random data: analysis and measurement procedures, 4th edn. Wiley, Hoboken. https://doi.org/10.1002/9781118032428.ch11
12. Proakis JG, Manolakis DG (2006) Digital signal processing: principles, algorithms, and applications, 4th edn. Pearson Prentice Hall, Upper Saddle River
13. Vetterli M, Kovacevic J, Goyal VK (2014) Foundations of signal processing, 3rd edn. Cambridge University Press, Cambridge
14. Owen M (2012) Practical signal processing. Cambridge University Press, Cambridge
15. Kay SM (2013) Fundamentals of statistical signal processing, Volume III: practical algorithm development, 1st edn. Prentice Hall, Westford
16. Smith S (2002) Digital signal processing: a practical guide for engineers and scientists, 1st edn. Newnes, Amsterdam
17. Manolakis DG, Ingle VK (2011) Applied digital signal processing: theory and practice, 1st edn. Cambridge University Press, Cambridge. https://doi.org/10.1017/cbo9780511835261
18. Aliev TA, Rzayev AH, Guluyev GA et al (2018) Robust technology and system for management of sucker rod pumping units in oil wells. Mech Syst Signal Process 99:47–56. https://doi.org/10.1016/j.ymssp.2017.06.010

Chapter 5
Application of Technology and System of Noise Control on Fixed Offshore Platforms and Drilling Rigs

Abstract Causes of numerous accidents on fixed offshore platform s are analyzed. It is revealed that the noise of noisy signals received by sensors of control systems in many cases contain valuable diagnostic information related to the beginning of platform's transition into the latent period of an emergency state. Taking into account this peculiarity, the structural principle of intelligent into an emergency state is proposed. It is also shown that most of the accidents on drilling rigs occur during drilling as a result of delayed measures on their prevention on the driller's part. To signal the beginning of the latent period of failures of drilling rigs, a subsystems of noise control of the beginning of the latent period of transition of offshore oil platforms technology and a toolkit for determining the moment of the appearance of a correlation between the useful signal and the noise is proposed, which will allow the driller to take timely measures to eliminate possible accident situations. This allows eliminating, to a great extent, the relationship between the occurrence of accidents and the driller's health, fatigue, and qualifications.

5.1 Systems for Noise Control of the Beginning of the Latent Period of Accidents on Fixed Offshore Platforms

Modern oil platforms are complex engineering facilities designed for well drilling producing hydrocarbons deposited on the bed of a body of water.

Circa 1891, the first submerged oil wells were drilled from platforms built on piles in the fresh waters of Grand Lake in the US state of Ohio [1, 2]. The wells were developed by small local companies such as Bryson, Riley Oil, German-American and Banker's Oil. The first offshore oil platform, Neft Dashlari (Oil Rocks), was built on metal piers in 1949 in the Caspian Sea 55 km from the coast of Azerbaijan. It is registered in the Guinness Book of Records as the oldest offshore oil platform [1, 3].

Different types of platforms are used, depending on water depth and other factors (Fig. 5.1) [1, 2]:

- Fixed platform
- Compliant tower

Fig. 5.1 1 and 2 are conventional fixed platforms; 3 is compliant tower; 4 and 5 are vertically moored tension leg and mini-tension-leg platform; 6 is spar; 7 and 8 are semi-submersibles; 9 is floating production, storage, and offloading facility; 10 is subsea completion and tie-back to host facility [4]

- Semi-submersible platform
- Jack-up drilling rig
- Drillship
- Floating production system
- Tension-leg platform
- Gravity-based structure
- Spar platform
- Normally unmanned installation (NUI)
- Conductor support system

A fixed platform (FP) is, from an engineering point of view, the simplest and most reliable and therefore the most common type of offshore platform. FPs are economically feasible for installation in water depths up to 500 m. FPs stand on steel or concrete legs anchored to the seabed. Drilling rigs, production equipment, crew quarters and auxiliary areas are installed on the deck in the upper body of the platform. FP is installed for long-term use—for instance, the world's largest oil platform, Hibernia platform (in the Atlantic Ocean, Canada, in 80 m of water, operating since 1997) [1, 5].

FPs are constantly subjected to environmental influence, such as waves, wind, temperature extremes, earthquakes, vibrations caused by landing helicopters, and shocks caused by mooring tankers and ships. These factors lead to the fatigue of metal structures, corrosion, microcracks, etc. An alteration in the technical condition of a platform's structure can eventually cause its catastrophic breakdown accompanied by material damage, casualties, and environmental pollution. To gain insight into the scale of this type of accident and the importance of finding a solution to this problem, we will briefly review the recent largest accidents on FPs.

The accident on an FP of the Bulla–Deniz offshore gas field on August 18, 2013 [1, 6, 7] and the accident on an FP of the Gunashli oilfield on December 4, 2015 [1,

5.1 Systems for Noise Control of the Beginning …

8–11] can serve as examples of such disasters in the Caspian Sea. Fire suppression during each of those accidents took more than 2 months, and the damages amounted to millions of dollars. The accident on the Gunashli platform, unfortunately, leads to the deaths of 29 oil workers. On November 6, 2014, an accident took place on an FP in the Caspian Sea, when a 250 mm pipeline on the pier leading to the platform collapsed and fell into the sea. Another accident occurred on a different FP on October 23, 2014, when a wagon-house (bunk house) fell into the sea, damaging and causing a fire on a 700 mm pipeline.

Similar accidents occurred at other oilfields as well [1]:

- On April 1, 2015, an explosion caused a fire on an FP in the Gulf of Mexico.
- On February 2015, an explosion occurred on an FP operated by the Brazilian energy producing company Petrobras. The platform was located near the Brazilian coast, northeast of the city of Vitoria.
- On November 21, 2014, news came about an accident on an FP operated by the Texan company Fieldwood Energy in the Gulf of Mexico. The explosion occurred 20 km off the coast of New Orleans.
- On December 28, 2013, a leak was discovered on a North Sea FP of the Norwegian company Statoil.
- On July 23, 2013, a fire broke out on a platform in the Gulf of Mexico approximately 88 km south of the coast of the US state of Louisiana.
- On January 16, 2013, TAQA Bratani (a subsidiary of Abu Dhabi National Energy) shut down the Brent System pipeline system in the British sector of the North Sea because of a leak that caused the shutdown of eight other platforms.
- On November 16, 2013, a fire broke out on an FP in the Gulf of Mexico off the coast of the US state Louisiana. The facility was operated by Black Elk Energy.
- On December 18, 2011, the Kolskaya drilling rig capsized and fully sank into the Sea of Okhotsk, 200 km off the coast of Sakhalin island.

Unfortunately, the frequency of accidents is not decreasing despite the fact that relevant systems are used for control and diagnostics of the technical condition of such platforms [12]. The best-known and most efficient systems for the control of the technical condition of FPs today are as follows:

- FUGRO Offshore Structural Monitoring Systems [13]. The complex includes an OLM System, which allows one to assess the structural integrity of an FP by monitoring the natural frequencies of structural components.
- BMT's Offshore Platform Monitoring Systems feature a Current Profiling System (CPS) [14] and Independent Remote Monitoring System (IRMS) [15] as well as a Subsea Strain Sensor Assembly (SSSA) [16] for controlling the structural integrity of subsea structures (tendons, production risers, steel catenary risers, platform legs, and braces). The feature of remote control allows one to continually monitor the condition of a platform in case an emergency evacuation of the operating personnel becomes necessary.

Unfortunately, the use of these systems notwithstanding, the accident rate for offshore oil and gas extracting facilities and communications, main oil and gas pipelines, etc., remains unreasonably high because of the shortcomings of their control and diagnostic systems.

An analysis of the causes of the described accidents shows that systems of FP control and diagnostics currently cannot control the beginning of the latent period of transition of an FP into an emergency state. Therefore, to lower the accident rate, we must develop new efficient technologies and systems for early detection of alterations in the technical condition of a platform. This would allow for timely evacuation of operating personnel and necessary repairs. Our analysis of numerous accidents has demonstrated that the transition of an FP into an emergency state starts with the initiation of various defects that emerge during the operation of a facility. The initiation of defects first shows in the noises of the signals received at the outputs of the sensors installed in the most vulnerable spots of the platform structure, which is why the noise becomes the carrier of the information about defect initiation. There are many effective methods of analysis of random noisy signals, such as spectral analysis, correlation analysis, wavelet analysis, regression analysis, etc. They are efficiently enough in the diagnostics of the technical condition of platforms for cases when the useful signal $X(i\Delta t)$ and the noise $\varepsilon(i\Delta t)$ have zero correlation. However, defect initiation in many cases manifests itself in the signals as noise that correlates with the useful signal. Therefore, when a defect is incipient, the presence of a correlation leads to errors in the estimates of the characteristics obtained by the mentioned methods of analysis of noisy signals [1]. Those errors make traditional methods unsuitable for identifying defects at their nascent stage. The control is performed when defects become strongly pronounced. For this reason, the control of defects by traditional methods often gives delayed results. Consequently, detection of a defect at its nascent stage requires calculating the estimates of characteristics of the noise and not the sum (noisy) signal. The advantage of this book is that it proposes exactly this kind of control. Thus, delayed results of control of the beginning of a platform's transition into an emergency state as performed by existing control systems sometimes lead to catastrophic consequences described earlier.

It is obvious that there are three possible ways to solve this problem:

1. Equipping existing control and diagnostics systems with technologies for detecting the beginning of the latent period of defect initiation and early warning of the beginning of accidents.
2. Equipping an already operating system with a subsystem for control of the beginning of the latent period of a platform's transition into an emergency state.
3. Designing an independent system for noise control of the beginning of the initiation of defects that lead to alteration of a platform's technical condition.

In the following paragraphs, we consider the third option, which can enhance the operating safety of FPs.

It follows from the above that the impossibility of extracting and accounting for the information contained in the noise of analyzed signals in existing systems of control for the FP technical condition is the main challenge in the control of the beginning of the latent period of their transition into an emergency state. This, in turn, complicates early forecasting of accidents at these facilities. Therefore, it is advisable to develop

5.1 Systems for Noise Control of the Beginning ...

noise technologies and systems capable of performing the control and identification by means of a matrix equivalent to platform's technical condition. We demonstrated the possibility of forming such matrices in Chap. 3.

In these matrices, the estimates of the errors $R_{X\varepsilon}(\mu \Delta t)$ are calculated from the expression

$$R_{X\varepsilon}(\mu \Delta t) \approx \frac{1}{2} R'_{X\varepsilon}(m \Delta t)$$

$$\approx \frac{1}{2N} \sum_{i=1}^{N} [g(i\Delta t)g((i+m)\Delta t) - 2g(i\Delta t)g((i+(m+1))\Delta t)$$

$$+ g(i\Delta t)g((i+(m+2))\Delta t)].$$

Through this, by calculating the estimates of D_ε, $R_{\varepsilon\varepsilon}(0)$, $R_{X\varepsilon}(0)$, $R_{X\varepsilon}(\Delta t)$, $R_{X\varepsilon}(2\Delta t)$, $R_{X\varepsilon}(3\Delta t)$, ... the corresponding elements of the correlation matrices are corrected. For instance, In the presence of a correlation between $X(i\Delta t)$ and $\varepsilon(i\Delta t)$, the elements of $R_{gg}(\Delta t)$, $R_{gg}(2\Delta t)$ and $R_{gg}(3\Delta t)$ are corrected by subtracting from them the respective estimates of $R_{X\varepsilon}(\Delta t)$, $R_{X\varepsilon}(2\Delta t)$ and $R_{X\varepsilon}(3\Delta t)$. For clarity, we demonstrate below the procedure of correction for the case when

$$R_{X\varepsilon}(0) > 0, \; R_{X\varepsilon}(\Delta t) > 0, \; R_{X\varepsilon}(2\Delta t) \approx 0, \; R_{X\varepsilon}(3\Delta t) \approx 0,$$

according to which, by using the estimates of D_ε, $R_{X\varepsilon}(0)$ and $R_{X\varepsilon}(\Delta t)$, the corresponding elements of the matrices $\bar{R}^e_{gg}(\mu \Delta t)$ and $\bar{r}^e_{gg}(\mu \Delta t)$ are corrected

$$\bar{R}^e_{gg}(\mu \Delta t) \approx \bar{R}_{XX}(\mu \Delta t),$$
$$\bar{r}^e_{gg}(\mu \Delta t) \approx \bar{r}_{XX}(\mu \Delta t),$$

It is obvious that after such a correction, the obtained matrices can be regarded as equivalent to the matrices of the useful signals.

It has been revealed in the process of research that a test of the effectiveness of the developed technology and control system by thorough experiments on real-life operating FPs takes a long time. For this reason, we built a test bench for simulating processes similar to those occurring on FP during their operation in real conditions. The test bench (Figs. 5.2 and 5.3) allowed us to conduct long-term seminatural experiments, simulating all types of extreme variants of platform operation. In corresponding vulnerable spots of the platform structure, we recreated such defects as microcracks, microde-formations, etc., and appropriate results were obtained with the use of the proposed technologies in the process of platform operation. Those results were compared with the results of similar experiments that we managed to conduct on actual platforms, and they almost always matched or were close, proving the expediency of the test bench in seminatural experiments.

Figure 5.4 shows the block diagram of the system of noise control of the beginning of the latent period of FP transition into an emergency state. In the system, the

Fig. 5.2 The appearance of the test bench simulating an FP for seminatural experiments

Fig. 5.3 A fragment of the test bench simulating an FP (with one BeanDevice AX-3D vibration sensor of the intelligent monitoring system in the center)

information on platform's technical condition affects the noisy signals received at the outputs of the Accutech AM20 acoustic sensors and BeanDevice AX-3D vibration sensors installed in all of platform's sections [1]. The measurement information is collected via Wi-Fi link by means of a BeanGateWay Controller. Acoustic and vibration signals are analyzed, informative attributes are formed and a knowledge base is created on a Getac A770 industrial computer. To enhance the certainty of the control results, the technology for calculating the estimates of the robust normalized correlation functions is combined with the calculation of the noise characteristics of acoustic

5.1 Systems for Noise Control of the Beginning …

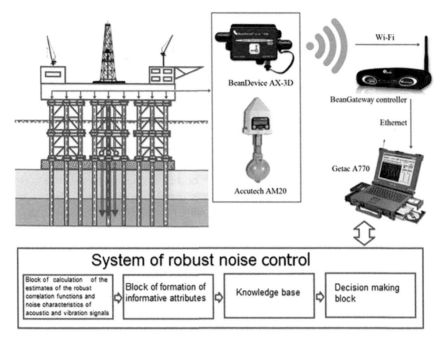

Fig. 5.4 Block diagram of the system of noise control of the beginning of the latent period of FP transition into an emergency state

and vibration signals. After many experiments, it was established that it is appropriate to sample acoustic and vibration signals at the frequency $f = 1000$–2000 Hz. With the amount of samples $U(i\Delta t)n \geq 4048$–8096, we obtained stable estimates of robust normalized correlation functions and noise characteristics.

During the experiments, it was also established that FP accidents are always preceded by the initiation of some defects. For instance, it was discovered that one of the most common causes of FP accidents is cracking caused by cyclic loads. As a rule, the process of crack development starts with the formation of microcracks, "coarsening" of the surface, grain boundary cracking, and cracking around solid inclusions. The process is accompanied by further infiltration deep into the material. Sometimes a microcrack becomes a macrocrack and spreads into the metal rather quickly. The speed of crack growth depends on the metal fatigue and the operating conditions. Our experiments also demonstrate that when microcracks form in metal trusses and supports of FPs, one first hears faint squeaking, which then grows louder; it then becomes reminiscent of a baby crying, turns into metallic creaking, etc. The spectra of the noise and the useful acoustic signal constantly change, as do the estimates D_ε, $R_{X\varepsilon}(\Delta t)$, $R_{X\varepsilon}(2\Delta t)$, …, $R_{X\varepsilon}(m\Delta t)$, which allows us to use them as carriers of information about defect initiation.

When a pinhole forms in an oil and gas pipeline connected to platform's compressor stations, one first hears faint whistling, which then grows louder, then the pipe starts humming and finally snorting and gurgling. The same happens to the noise of the signal received at the output of the vibration sensor. First, a faint high-frequency noise appears, then the spectrum of the noise decreases, the amplitude increases. Therefore, the characteristics of the noise of the signals received at the outputs of acoustic and vibration sensors change continuously this entire time as well. It is obvious that the estimates of the noise characteristics can also be taken as informative attributes of the beginning of the latent period of emergence of a pinhole that can afterwards cause an accident.

When a different defect starts emerging, such as weakening of the fixing point of the structural channel on the truss of a platform, a high-frequency noise also appears in vibration and acoustic sensors of the corresponding sensors. As this defect develops, the frequency of the noise spectrum gradually decreases, and this process continues until the platforms go into an emergency state. In that case, the estimates of normalized correlation functions and noise characteristics also become informative attributes of the beginning of the latent period of an emergency state.

Thus, we have established by way of experiments that the initiation of any of the described defects affects the estimates of the normalized correlation functions and the noise characteristics of acoustic and vibration signals received from the sensors of corresponding sections of FP structure. Our analysis of the noisy signals received from the acoustic and vibration sensors on FP also demonstrated that the initiation of various defects leads to the emergence of a correlation between $X(i\Delta t)$ and $\varepsilon(i\Delta t)$. Moreover, the experiments have also revealed that the time shifts increase depending on the dynamics of defect development, with the maximum time shift not exceeding $\mu\Delta t = 6\Delta t$, i.e., the correlation disappeared at $\mu\Delta t = 6\Delta t$ (in our case, $\Delta t \approx 1$ ms).

During the operation of the system described in Fig. 5.4, a combination of the mentioned informative attributes forms. Naturally, it occurs during the operation of an FP in both normal and extreme weather conditions. When microchanges in the technical condition of an FP start, the block of formation of informative attributes also forms corresponding reference combinations of the robust estimates of the correlation functions and the estimates of the noise characteristics. This occurs when all other possible faults start to appear. At that time, the sought-for estimates change, and the corresponding reference combinations of vibration and acoustic signals form. A set of certain combinations of those estimates corresponds to each of the possible faults. As a result, after a certain operation period, due to the initiation of various defects, sets of reference combinations are formed and saved in the knowledge base both for the normal state and for the corresponding faults.

Thus, the system in Fig. 5.4 controlling the beginning of the latent period of facility's transition into an emergency state first operates in the training mode. During this time, first of all, the reference informative attributes are calculated and the corresponding sets form. To improve the reliability and validity of the control results, signal processing is duplicated with the use of various signal analysis technologies. As a result, the corresponding reference combinations of informative attributes are compiled from the estimates of the noise characteristics of every noisy acoustic and vibration signal in the following form:

5.1 Systems for Noise Control of the Beginning ...

$$W_{g_1} = \begin{cases} D_{g_1\varepsilon}, R_{g_1 X\varepsilon}(1\Delta t), R_{g_1 X\varepsilon}(2\Delta t), R_{g_1 X\varepsilon}(3\Delta t), \ldots, R_{g_1 X\varepsilon}(m\Delta t) \\ R^*_{g_1 X\varepsilon}(1\Delta t), R^*_{g_1 X\varepsilon}(2\Delta t), R^*_{g_1 X\varepsilon}(3\Delta t), \ldots, R^*_{g_1 X\varepsilon}(m\Delta t) \\ a^R_{g_1 n}, b^R_{g_1 n}; a_{g_1 n\varepsilon}, b_{g_1 n\varepsilon}; a_{g_1 n\varepsilon_2}, b_{g_1 n\varepsilon_2} \\ a^*_{g_1 n}, b^*_{g_1 n}; a^*_{g_1 n\varepsilon}, b^*_{g_1 n\varepsilon}; a^*_{\underline{g_1} n\varepsilon_2}, b^*_{\underline{g_1} n\varepsilon_2} \\ k_{g_1 q_0}, k_{g_1 q_1}, k_{g_1 q_2}, \ldots, k_{g_1 q_m}; f_{g_1 q_0}, f_{g_1 q_1}, f_{g_1 q_2}, \ldots, \bar{f}_{g_1 q_m} \end{cases}$$

$$W_{g_2} = \begin{cases} D_{g_2\varepsilon}, R_{g_2 X\varepsilon}(1\Delta t), R_{g_2 X\varepsilon}(2\Delta t), R_{g_2 X\varepsilon}(3\Delta t), \ldots, R_{g_2 X\varepsilon}(m\Delta t) \\ R^*_{g_2 X\varepsilon}(1\Delta t), R^*_{g_2 X\varepsilon}(2\Delta t), R^*_{g_2 X\varepsilon}(3\Delta t), \ldots, R^*_{g_2 X\varepsilon}(m\Delta t) \\ a^R_{g_2 n}, b^R_{g_2 n}; a_{g_2 n\varepsilon}, b_{g_2 n\varepsilon}; a_{g_2 n\varepsilon_2}, b_{g_2 n\varepsilon_2} \\ a^*_{g_2 n}, b^*_{g_2 n}; a^*_{g_2 n\varepsilon}, b^*_{g_2 n\varepsilon}; a^*_{\underline{g_2} n\varepsilon_2}, b^*_{\underline{g_2} n\varepsilon_2} \\ k_{g_2 q_0}, k_{g_2 q_1}, k_{g_2 q_2}, \ldots, k_{g_2 q_m}; f_{g_2 q_0}, f_{g_2 q_1}, f_{g_2 q_2}, \ldots, \bar{f}_{g_2 q_m} \end{cases}$$

$$W_{g_j} = \begin{cases} D_{g_j\varepsilon}, R_{g_j X\varepsilon}(1\Delta t), R_{g_j X\varepsilon}(2\Delta t), R_{g_j X\varepsilon}(3\Delta t), \ldots, R_{g_j X\varepsilon}(m\Delta t) \\ R^*_{g_j X\varepsilon}(1\Delta t), R^*_{g_j X\varepsilon}(2\Delta t), R^*_{g_j X\varepsilon}(3\Delta t), \ldots, R^*_{g_j X\varepsilon}(m\Delta t) \\ a^R_{g_j n}, b^R_{g_j n}; a_{g_j n\varepsilon}, b_{g_j n\varepsilon}; a_{g_j n\varepsilon_2}, b_{g_j n\varepsilon_2} \\ a^*_{g_j n}, b^*_{g_j n}; a^*_{g_j n\varepsilon}, b^*_{g_j n\varepsilon}; a^*_{g_j n\varepsilon_2}, b^*_{g_j n\varepsilon_2} \\ k_{g_j q_0}, k_{g_j q_1}, k_{g_j q_2}, \ldots, k_{g_j q_m}; f_{g_j q_0}, f_{g_j q_1}, f_{g_j q_2}, \ldots, \bar{f}_{g_j q_m} \end{cases}$$

After that, these combinations of reference informative attributes form the sets of reference informative attributes (SRIA) W_j that correspond to facility's normal technical condition.

The duration of the training process depends on the specifics of weather conditions. In the future, after the training is complete, if in extreme weather conditions any combination of estimates matches the one already present in SRIA, the decision-making block registers the fact of the match. If the current combination differs from all those present in SRIA, a new reference is entered in the corresponding set. At some point, newly obtained current estimates repeatedly turn out to be equal or close in value to certain combinations in the corresponding set in SRIA. The first stage of the training process is regarded as complete only after such results have been obtained on multiple occasions. Thus, at the first stage, SRIA forms, which allows us to perform the control of the beginning of accidents.

At the second stage, the monitoring of the technical condition of a platform starts. The calculation of the informative attributes and the formation of combinations from them is carried out in the similar way. However, in this case, if a current combination of estimates matches the estimates already present in SRIA from the set corresponding to the normal condition, this is taken as the confirmation of the normal technical condition. If a current combination in some time period differs from all sets corresponding to the normal condition and matches the combination corresponding to a fault already present in SRIA, then information about possible microchanges in the technical condition in the corresponding section of FP is formed and displayed on the monitor of the operating personnel. The personnel is also provided with the list of all combinations of estimates from SRIA that correspond to that condition, as well as the estimates from the current combination. If no malfunction is detected after appropriate measures of diagnostics and control of facility's technical condition, and it is established that the facility operates normally, that current combination is also included in the corresponding set of reference combinations and saved in SRIA. Otherwise, the beginning of the transition of the platform into an emergency state is recorded, the appropriate information is formed, which is registered as a report indicating the time and results of monitoring.

The algorithms and technologies presented in Chaps. 2–4 make it possible to form correlation matrices of noisy random processes equivalent to the matrices of their useful signals both in the absence of a correlation between the useful signal and the noise and in the presence of such. Thus, we eliminate the difficulties of solving the problems of identifying the technical condition of FP with the use of correlation matrices. This, in turn, allows us to perform control of platforms for possible accident situations, detecting them at their initial stages, making it possible to take timely measures for accident prevention. However, our experiments have shown that solving the problems of identifying and diagnosing platform's technical condition based on the matrix equation with the use of the correlation matrices given in Chap. 3 takes a long time. Therefore, to solve this problem in real time, it is advisable to carry out noise control of the beginning of initiation of various defects that subsequently lead the platform to an emergency state. According to the results of our seminatural experiments, the use of the noise technology in the proposed systems in the above variant makes it possible to obtain results of control of the beginning of initiation of latent defects in real-life facilities from several days to several months earlier than it is done by traditional technologies. This is because the emergence of various defects, such as wear and tear, microcracks, and fatigue strain in the process of platform operation, lead to changes in its technical condition. Such changes manifests themselves as the sounds $\varepsilon_{21}(i\Delta t)$, $\varepsilon_{22}(i\Delta t)$, $\varepsilon_{23}(i\Delta t)$, ..., $\varepsilon_{2m}(i\Delta t)$ in the noises $\varepsilon_1(i\Delta t)$, $\varepsilon_2(i\Delta t)$, $\varepsilon_3(i\Delta t)$, ..., $\varepsilon_m(i\Delta t)$ of the signals $g_1(i\Delta t)$, $g_2(i\Delta t)$, $g_3(i\Delta t)$, ..., $g_m(i\Delta t)$, and this, in turn, affects the aforementioned estimates of these signals. Our numerous experiments have confirmed the effectiveness of this technology and demonstrated the expediency of its practical application for the control of the beginning and development dynamics of accidents on fixed offshore platforms.

5.2 Technologies and System for the Noise Control of the Beginning and Dynamics of the Development of Accidents on Drilling Rigs

It is known that drilling accidents lead to environmental pollution of soil, nearby rivers and reservoirs as well as to substantial material damage. Therefore, conducting research related to the development of new effective technologies and systems for improving the safety of the drilling process is an extremely urgent task [17].

Currently, the most common type of drilling is rotary drilling, in which the rock-cutting tool receives rotation from a special mechanism (spindle rotator or rotor) through a string of drill pipes or from a downhole motor.

Various modern systems are developed and used for diagnosing and managing the drilling process to minimize the occurrence of possible accidents, All these drilling control and management systems perform sensor readings in real time, perform measurement processing and continuously control and manage the full technological cycle of well construction, and perform predictions for timely prevention of accident situations [17–20].

The set of parameters to be controlled when drilling deep wells includes the hook weight, pressure of the drilling fluid at the well inlet, density of the drilling fluid at the well inlet, rotor torque, flow rate of the drilling fluid at the well outlet, flow rate of the drilling fluid at the well inlet, speed of round-trip operations, drilling penetration rate, and outlet temperature [17–20].

Despite the use of the above systems, the process of drilling is currently accompanied by an unjustifiably large number of costly accidents. The emergence of accidents during drilling is attributed to specific factors of accidents, such as the multifactority and uncertainty of the mechanisms of occurrence of accidents their regional specificity, speed, inaccessibility for instrumental control, and vagueness and ambiguity of observed symptoms [17–20]. The results of the measurements conducted during the drilling process are influenced by random impacts that change the actual conditions of the operation of the equipment during the measurement. In addition, the technological process of well drilling is characterized by the following: a large number of random factors that change over time and affect the quality and technical and economic performance of the operation, the variety of geological and technical conditions of drilling, as well as the distortion of the useful signal (the hook load, torque, power expenditure, rate of penetration, etc.) that is used to determine the parameters of the drilling mode [17]. Consequently, the main technological parameters of drilling are random functions. For the above reasons, the algorithms and technologies used in practice do not always provide timely warning of possible accidents during well drilling and must be improved. In this regard, many specialists reasonably believe that the efficiency and safety of drilling when using existing systems depend to a significant extent on the driller's skills. In addition, this process affects the driller's health, fatigue, fatigue, and even mood. For instance, according to the accident time statistics, many catastrophes at drilling rigs occurred at night, when the driller was sleep-deprived and tired. Therefore, it is necessary to create new technologies and tools that, by identifying the beginning of the latent period of accidents and alerting the driller to it, will focus the driller's attention on the threat. It will also be possible to ensure reliability and safety of the drilling process regardless of the driller's skill and state of health.

As indicated above, the common practice these days is rotary drilling, in which the rock-cutting tool is rotated by a special mechanism, a spindle rotator or rotor through a drill string, or by a downhole motor [17]. All these processes inevitably affect the signals received from the sensors of such controllable drilling parameters as the bit rotation frequency $g_1(t)$, the torque at the spindle of the swivel head $g_2(t)$, the torque on the rotor of the drilling rig $g_3(t)$, the rate of penetration $g_4(t)$ and the axial load on the drilling bit $g_5(t)$. They carry certain information regarding the technical condition of the drilling machine. Therefore, the informative attributes generated from the noises of these signals can be used for determining the beginning of the latent period of accident initiation. This is due to the fact that by determining the moment of appearance of the correlation between the useful signal and the noise of these signals, one can adequately control the beginning of the change in the technical condition of the drilling machine. Therefore, it is possible to create software tools for the driller, which by detecting the moments of the appearance of a correlation between the corresponding useful signals and noises will signal the beginning of the latent period of initiation of accident situation. Our experimental studies have shown that those processes that cause deviations from the normal drilling mode also indirectly affect the vibration state of such major parts of the drilling rig as the rotary table, the top of the drilling rig, the middle of the machine frame, etc. Highly skilled experienced drillers determine the beginning of possible accident practically unmistakably, by touch, by the vibration state of these parts of the drilling rig. Therefore, to control the onset of an accident situation, it is also possible to use the noises from the vibration signals received from the sensors of the drilling rig installed on these drilling rig components. In this variant informative attributes consisting of the estimates of the noises of the signals received from vibration sensors VP1–VP3 installed on said components are used to control the onset of the latent period of an emergency state. Note that there is a third alternative, when the above-described variants are combined and thus greater reliability and validity of the results of determining a possible accident situation.

Taking this into account, let us consider the second variant solution to the problem of developing a toolkit that allows monitoring the beginning of accidents by analyzing the noise of the signals received from the corresponding components of the drilling rig. In this case, diagnostic information from the noises $\varepsilon_{V1}(t)$, $\varepsilon_{V2}(t)$, $\varepsilon_{V3}(t)$ of the vibration signals $g_{V1}(t) = x_{V1}(t) + \varepsilon_{V1}(t)$, $g_{V2}(t) = x_{V2}(t) + \varepsilon_{V2}(t)$, $g_{V3}(t) = x_{V3}(t) + \varepsilon_{V3}(t)$ obtained from the sensors VP1–VP3 is used to control the onset of the latent period of an emergency state of the drilling process.

It is known that the following equality is valid in the normal state of the process for centered noisy vibration signals $g(t) = X(t) + \varepsilon(t)$ obtained at the outputs of the said sensor [17]:

$$D_g = M[g(t)g(t)] = M[(X(t) + \varepsilon(t))(X(t) + \varepsilon(t))]$$
$$= M[X(t)X(t) + \varepsilon(t)X(t) + X(t)\varepsilon(t) + \varepsilon(t)\varepsilon(t)], \quad (5.1)$$

where D_g is the variance of the sum signal $g(t)$.

In this case, in the absence of correlation between the useful signal $X(t)$ and the noise $\varepsilon(t) = \varepsilon_1(t)$ the following conditions are fulfilled:

$$M[X(t)X(t)] \neq 0, \quad M[\varepsilon_1(t)X(t)] = 0,$$
$$M[X(t)\varepsilon_1(t)] = 0, \quad M[\varepsilon_1(t)\varepsilon_1(t)] \neq 0$$

5.2 Technologies and System for the Noise Control ...

$$D_g = M[X(t)X(t) + \varepsilon_1(t)\varepsilon_1(t)]. \qquad (5.2)$$

Therefore, in this case, we have

$$D_g = M[X(t)X(t) + \varepsilon_1(t)\varepsilon_1(t)] = R_{XX}(0) + D_{\varepsilon_1},$$

where

$$D_{\varepsilon_1} = M[\varepsilon_1(t)\varepsilon_1(t)] = M[\varepsilon(t)\varepsilon(t)]. \qquad (5.3)$$

At the beginning of the latent period of an emergency state, along with the noise $\varepsilon_1(t)$, the noise $\varepsilon_2(t)$ correlated with and useful signal $X(t)$ that carries the information on the change in the technical condition of the facility appears in the signals $g(t)$ on the rig [17]. At the same time, due to the presence of a correlation between the useful signal $X(t)$ and the sum noise $\varepsilon(t) = \varepsilon_1(t) + \varepsilon_2(t)$, the fulfillment of conditions (5.1)–(5.3) is violated because the following is valid:

$$\begin{aligned}D_g &= M[(X(t) + \varepsilon_1(t) + \varepsilon_2(t))(X(t) + \varepsilon_1(t) + \varepsilon_2(t))] \\ &= M[X(t)X(t)] + M[X(t)\varepsilon_1(t)] + M[X(t)\varepsilon_2(t)] \\ &+ M[\varepsilon_1(t)X(t)] + M[\varepsilon_1(t)\varepsilon_1(t)] + M[\varepsilon_1(t)\varepsilon_2(t)] \\ &+ M[\varepsilon_2(t)X(t)] + M[\varepsilon_2(t)\varepsilon_1(t)] + M[\varepsilon_2(t)\varepsilon_2(t)].\end{aligned}$$

Taking into account that

$$\begin{aligned}&M[X(t)\varepsilon_1(t)] = 0, \quad M[\varepsilon_1(t)X(t)] = 0, \\ &M[\varepsilon_1(t)\varepsilon_2(t)] = 0, \quad M[\varepsilon_2(t)\varepsilon_1(t)] = 0, \\ &M[\varepsilon_1(t)\varepsilon_1(t)] + M[\varepsilon_2(t)\varepsilon_2(t)] = D_{\varepsilon_1} + D_{\varepsilon_2} = D_\varepsilon,\end{aligned}$$

we have

$$\begin{aligned}D_g &= M[X(t)X(t)] + M[X(t)\varepsilon_2(t)] + M[\varepsilon_2(t)X(t)] + D_{\varepsilon_1\varepsilon_1} + D_{\varepsilon_2\varepsilon_2} \\ &= D_X + 2M[X(t)\varepsilon_2(t)] + D_{\varepsilon\varepsilon}.\end{aligned}$$

In view of the above, in order to solve the problem of the noise control of the beginning and development dynamics of the latent period of drilling accidents, it is necessary to ensure the extraction of diagnostic information from the noise $\varepsilon_2(t)$.

Studies [17] have demonstrated that the beginning of the latent period of an emergency state of drilling rigs in the presence of a correlation between the noise and the useful signal, before affecting the estimates of the spectral characteristics of vibration signals, first manifests itself in the spectrum of the noise $\varepsilon(i\Delta t)$. Therefore, when solving the problem under consideration, it is appropriate to use as informative attributes the estimates of the spectral characteristics of the noise $\varepsilon(i\Delta t)$ vibration signals, which can be calculated from expressions

$$a_{n_\varepsilon} \approx \frac{2}{N} \sum_{i=1}^{N} \text{sgn}\varepsilon'(i\Delta t)\sqrt{|\varepsilon'(i\Delta t)|}\cos n\omega(i\Delta t),$$

$$b_{n_\varepsilon} \approx \frac{2}{N} \sum_{i=1}^{N} \text{sgn}\varepsilon'(i\Delta t)\sqrt{|\varepsilon'(i\Delta t)|}\sin n\omega(i\Delta t),$$

$$a_{n_\varepsilon}^* \approx \frac{2}{N} \sum_{i=1}^{N} \text{sgn}\varepsilon'(i\Delta t)\cos n\omega(i\Delta t),$$

$$b_{n_\varepsilon}^* \approx \sum_{i=1}^{N} \text{sgn}\varepsilon'(i\Delta t)\sin n\omega(i\Delta t).$$

Because it is extremely important to ensure the reliability and validity of the results of control of the onset of the latent period of faults on drilling rigs, it is expedient to duplicate these algorithms with the technologies of correlation analysis of the noise. An experimental analysis of the most effective variants showed that for this purpose, it is appropriate to use the algorithm for calculating the estimate of the relay cross-correlation function $R_{X\varepsilon}^*(\mu)$ between $X(i\Delta t)$ and $\varepsilon(i\Delta t)$ of the vibration signal $g(i\Delta t)$.

$$R_{X\varepsilon}^*(0) = \frac{1}{N} \sum_{i=1}^{N} \text{sgn} g(i\Delta t)[g(i\Delta t) + g((i+2)\Delta t) - 2g(i+1)\Delta t]. \quad (5.8)$$

The appeal of this algorithm is that in the normal state of the drilling rig, the estimate $R_{X\varepsilon}^*(0)$ is always zero. However, when different faults arise, such as a correlation between $X(i\Delta t)$ and $\varepsilon(i\Delta t)$, the estimate $R_{X\varepsilon}^*(0)$ differs from zero, which reliably indicates the beginning of the latent period of faults.

As was mentioned in the previous paragraphs, the drilling process these days is controlled by a driller through a rig control system (RCS). By means of this system, the drill operator can obtain all types of information in a timely manner and quickly manage the drilling process. However, in the initial latent period of an emergency state, the RCS does not provide the drill operator with adequate information; as a result, that moment is established by the driller intuitively. Thus, the probability of an accident to a certain extent depends on his/her qualifications. Therefore, to avoid the drill operator's possible mistakes, it is necessary to provide the driller with tools that facilitate her/his intuitive activity.

An experimental version of the subsystem of noise control of the onset and dynamics of the development of faults was created for this purpose. The diagram of the subsystem is shown in Fig. 5.5. In addition, to conduct seminatural experiments, a working model of the drilling rig was also constructed; a photograph of the model is shown in Fig. 5.6. As seen from the diagram in Fig. 5.5, the system consists of

5.2 Technologies and System for the Noise Control …

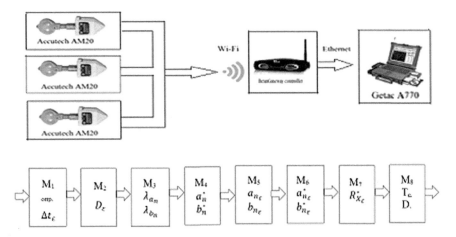

Fig. 5.5 Noise control subsystem

Fig. 5.6 Mock-up drilling rig for conducting seminatural experiments

three DV_1–DV_3 type Accutech AM20 vibration sensors, a BeanGateway controller type receiving antenna and a Getac A770 type computer. The DV_1–DV_3 sensors are mounted (attached) on the rotary table, on the top of the drilling rig and on the middle of the machine frame. These sensors measure and transmit vibration signals $gV_1(t)$, $gV_2(t)$, and $gV_3(t)$ by radio channel at a distance of up to 650 meters. The BeanGateway controller receives these signals and then enters them into the computer.

In the process of drilling, torsional vibration, axial vibration, and lateral vibration cause the sum vibrational process which, by means of DV_1–DV_3 sensors, is transformed into noisy vibration signals, $gV_1(i\Delta t)$, $gV_2(i\Delta t)$ and $gV_3(i\Delta t)$. The signals are transmitted via a radio channel to a computer where, using the software modules M_1–M_8 and the corresponding algorithms, the combination of informative attributes are determined, consisting of the estimates of $D_\varepsilon; a_n^*, b_n^*; a_{n_\varepsilon}, b_{n_\varepsilon}; a_{n_\varepsilon}^*, b_{n_\varepsilon}^*, \ldots, R_{x\varepsilon}^*(\mu)$ that indicate the technical condition of the rig. In module M_8, when the set threshold value of the estimates of these informative attributes is exceeded, the beginning T_C of the latent period of an emergency state is registered and the development dynamics D_p of the accident is determined.

Note that in addition to these estimates, to improve the monitoring results, analogous estimates obtained through an analysis of the signals of the bit rotation speed $g_1(i\Delta t)$, the torque on the spindle $g_2(i\Delta t)$, the torque on the rotor $g_3(i\Delta t)$, the rate of penetration $g_4(i\Delta t)$ and the axial load $g_5(i\Delta t)$ can also be used as informative attributes.

When the noise system is functioning during the drilling process, the vibration signals $g_{V1}(i\Delta t)$, $g_{V2}(i\Delta t)$ and $g_{V3}(i\Delta t)$ are sampled at intervals based on the frequency of the spectrum of the noise $\varepsilon(i\Delta t)$ (the technology for calculating the sampling interval Δt_ε is described in the following paragraphs). As a result, the estimates of informative attributes $D_\varepsilon; a_n^*, b_n^*; a_{n_\varepsilon}, b_{n_\varepsilon};$ and $a_{n_\varepsilon}^*, b_{n_\varepsilon}^*, \ldots, R_{X\varepsilon}^*(\mu)$ are calculated in real time. Our experiments have shown that during the drilling process, in periods when there are no accident situations, even if the estimates of the spectral and correlation characteristics of the sum noisy vibrating signals $g_{V1}(i\Delta t)$, $g_{V2}(i\Delta t)$ and $g_{V3}(i\Delta t)$ change within a wide range, the values of their noise characteristics are zero. However, from the onset and development of malfunctions in the drilling process, these estimates differ from zero, and as they develop, the values of these estimates also increase. If adverse processes stabilize, then the change in these estimates in time ceases. These specifics of the noise estimates of the vibration signal characteristics allow us to determine the beginning of the latent period of an emergency state of the drilling process and to control its development dynamics. This makes it possible to provide information to the driller, allowing her/him to make the optimal decision by choosing the most advantageous time for taking the necessary measures to prevent accidents.

As shown by the results of numerous experiments, when the dynamic of changes in the values of the estimates of the characteristics is slow or moderate, then it is advisable that the driller should wait until the information regarding the emergency state of the process appears on the RCS and then take appropriate measures. However, in those cases when there is an accelerated increase in the values of the noise estimates of the vibration signal characteristics, i.e., when the dynamic of the development of the accident process is high, the driller must take emergency measures to prevent an accident. At the same time, when the dynamic of the increase in the estimates of the noise characteristics stops, the drilling process can be continued in the normal mode. Presented in Fig. 5.7 are the results of one of our numerous experiments, showing three cases. Note that at the beginning of an accident situation, the combinations of signal estimates received from all three DV_1–DV_3 sensors are sent to the monitor screen of the noise system. These combinations are reflected on the screen in the first rows of the three respective tables. Tables 5.1, 5.2 and 5.3 show the results of only one sensor in the indicated three accident situations.

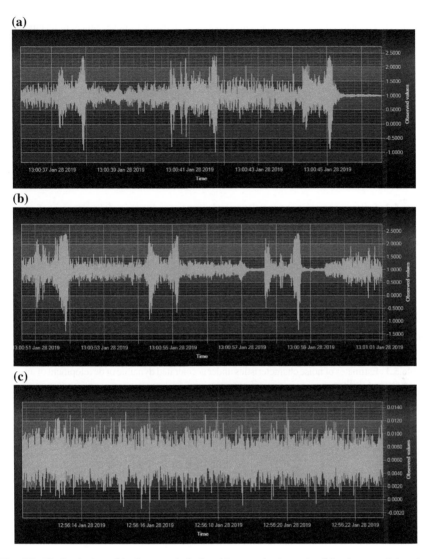

Fig. 5.7 The beginning of the latent period of accidents: **a** the dynamic of development is low (see Table 5.1); **b** the dynamic of development is moderate (see Table 5.2); **c** the dynamic of development is accelerated (see Table 5.3)

Table 5.1 Estimates of noise characteristics under low dynamic of development of an emergency state of drilling

D_ε	$R_{x\varepsilon}(0)$	$R_{x\varepsilon}(\Delta t)$	$R_{x\varepsilon}(2\Delta t)$	a_{1q}	b_{1q}	a_{2q}	b_{2q}	a_{3q}	b_{3q}
0.21	0.15	0.13	0.11	0.14	0.13	0.00	0.00	0.00	0.00
0.25	0.19	0.15	0.11	0.17	0.17	0.05	0.06	0.05	0.07
0.24	0.20	0.18	0.12	0.18	0.16	0.10	0.05	0.07	0.07
0.24	0.20	0.17	0.12	0.18	0.17	0.10	0.06	0.06	0.07
0.25	0.20	0.18	0.19	0.16	0.14	0.08	0.09	0.03	0.04
0.24	0.21	0.29	0.16	0.15	0.15	0.11	0.09	0.04	0.05

Table 5.2 Estimates of noise characteristics under moderate dynamic of development of an emergency state of drilling

D_ε	$R_{x\varepsilon}(0)$	$R_{x\varepsilon}(\Delta t)$	$R_{x\varepsilon}(2\Delta t)$	a_{1q}	b_{1q}	a_{2q}	b_{2q}	a_{3q}	b_{3q}
0.33	0.25	0.19	0.09	0.17	0.16	0.11	0.13	0.09	0.07
0.40	0.35	0.24	0.16	0.21	0.19	0.19	0.13	0.15	0.18
0.47	0.41	0.32	0.22	0.25	0.21	0.21	0.17	0.19	0.18
0.51	0.47	0.36	0.28	0.28	0.27	0.28	0.22	0.24	0.22
0.63	0.56	0.46	0.39	0.36	0.37	0.36	0.29	0.30	0.29

Table 5.3 Estimates of noise characteristics under accelerated dynamic of development of an emergency state of drilling

D_ε	$R_{x\varepsilon}(0)$	$R_{x\varepsilon}(\Delta t)$	$R_{x\varepsilon}(2\Delta t)$	a_{1q}	b_{1q}	a_{2q}	b_{2q}	a_{3q}	b_{3q}
0.12	0.06	0.03	0.00	0.06	0.02	0.01	0.00	0.00	0.05
0.28	0.19	0.15	0.11	0.19	0.11	0.09	0.07	0.12	0.11
0.41	0.36	0.22	0.24	0.19	0.28	0.11	0.14	0.21	0.19
0.64	0.59	0.38	0.37	0.34	0.28	0.21	0.28	0.29	0.29
0.77	0.73	0.54	0.09	0.39	0.36	0.32	0.34	0.35	0.36
0.85	0.81	0.73	0.64	0.43	0.41	0.37	0.36	0.41	0.44

The system is designed in such a manner that after the appearance of the first row, i.e., after receiving the warning information, the driller determines the time interval ΔT, which she/he considers sufficient to control the dynamic of the development of the accident process. In this case, the driller can consistently, row by row, obtain information with the interval $\Delta T = 2, \Delta T = 4, \Delta T = 5, \Delta T = 6$ and $\Delta T = 7$ min. After the appropriate time ΔT is set, the system gives the driller the next row in all three tables at the interval ΔT indicating the dynamic of the increase in the estimates compared to the previous row, in percentage terms. Thus, the driller receives the information about the beginning of the latent period of an accident process and the dynamics of its development immediately when the process starts.

5.2 Technologies and System for the Noise Control ...

Tables 5.1, 5.2 and 5.3 show the estimates of the noise characteristics of the vibration signal obtained from the DV_1 sensor, reflecting the three most typical variants. Tables 5.1, 5.2, and 5.3 show the beginning and development dynamics of accident situations. The results of numerous experiments repeated with slight differences.

According to Tables 5.1, 5.2 and 5.3, in the first variant, the dynamic of the development of an emergency state of drilling is low; in the second variant, the dynamic of development is moderate; and in the third variant, the dynamic of development is accelerated. In the first and second cases, the driller can plan a set of measures in advance to eliminate the malfunction. The driller can postpone the execution of those measures until the information on the emergency state is confirmed on the RCS. In the second variant, in cases when the driller does not want to take risks, she/he can begin to eliminate the causes of the accident, without waiting for the corresponding information from the RCS. However, in the third case, because of the accelerated development dynamic of the accident process, the driller must take immediate action, as waiting for the confirmation of the reliability of this information from the RCS can lead to delayed results, possibly resulting in an accident.

Thus, the proposed subsystem of noise control of the beginning of the latent period of the emergency state can greatly facilitate the driller's work in the event of ambiguous situations when the information obtained from traditional control systems is insufficient to take adequate measures for early accident prevention.

Our experimental studies have also shown that effective operation of the proposed tools requires algorithms and technologies for determining the sampling interval Δt_ε of the noise in real time. Therefore, we must dwell on this issue in more detail.

As indicated in Chaps. 1 and 2, frequencies of vibration processes in different equipment during the drilling period vary in a wide range and depends on many factors. Therefore, taking into account the time variation of the spectrum of the vibration signals $g_V(i\Delta t)$ and the noise $\varepsilon(i\Delta t)$ under the influence of these factors, the sampling interval is to be determined adaptively in the proposed system to obtain the required estimates with the required accuracy in real time. As a result, the sought-for estimates of the spectral characteristics of the noise $\varepsilon(i\Delta t)$ and the sum signal are calculated with sufficient accuracy [17].

This calculation is achieved by using the frequency characteristics of the least significant bit $q_0(i\Delta t)$ of the samples $g_V(i\Delta t)$ that is generated by the analogue-to-digital conversion of the analyzed vibration signals $g_{V1}(i\Delta t)-g_{V3}(i\Delta t)$. To perform this conversion, these characteristics are converted into a digital code with the frequency f_v, which significantly exceeds the necessary sampling frequency f_c found by the traditional method. Here, the following inequality is valid:

$$f_v \gg f_c.$$

During the analogue-to-digital conversion of vibration signals with the frequency f_v, the number N_{q_0} of the transitions of the least significant bit $q_0(i\Delta t)$ of the sample $g_V(i\Delta t)$ from the unit to the zero state in unit time, and the total number of samples of the vibration signal $g(i\Delta t)$ are calculated, and the formula

$$f_{q_0} \approx \frac{N_{q_0}}{N} f_v$$

is used to calculate the frequency f_{q_0} of the least significant digit $q_0(i\Delta t)$, which reflects the spectrum of the noise $\varepsilon(i\Delta t)$.

In the noise control subsystem, this calculation is implemented in the following way:

(1) the vibration signal $g_V(i\Delta t)$ during the observation time T is converted into digital code with the excess frequency f_v, and then a file is generated from N of its samples;
(2) the number of samples N_{q_0}, at which the least significant bit $q_0(i\Delta t)$ of sample $g_V(i\Delta t)$ goes from the unit state into the zero state is determined by software calculation;
(3) using the correlations

$$f_{q_0} = \frac{N_{q_0}}{N} f_v; \quad \Delta t_\varepsilon = \frac{1}{f_{q_0}}$$

f_{q_0} and Δt_ε are calculated.

For example, if at the sampling frequency $f = 10000$ Hz the number N_{q_0} is equal to 1000 samples, then the frequency $f_{q_0} = \frac{N_{q_0}}{N} f_v = \frac{1000}{10000} 10000 = 1000$ Hz and the interval Δt will be $\Delta t_\varepsilon \leq \frac{1}{f_{q_0}} = \frac{1}{1000} = 0.001$ s.

Our studies have shown that such calculation of the sampling interval Δt_ε is easily software-programmable, and during the encoding of the vibration signals $g(i\Delta t)$, it is easy to determine the sampling interval that corresponds to the spectrum of the noise $\varepsilon(i\Delta t)$ of the sum signal $g_V(i\Delta t)$. It has been established that at the sampling frequency of $f_v = 10000$–5000 Hz, the result of spectral noise monitoring of the vibration state can be considered effective.

As indicated above, the information regarding the beginning of the latent period of malfunctions in the drilling process received from DV_1–DV_3 vibration sensors largely depends on their technical characteristics and operating conditions.

It is well known that vibration control systems are usually equipped with sensors of vibration displacement, vibration velocity, and vibration acceleration.

Vibration displacement sensors are used to control an object's position, vibration velocity sensors, the rate of change of its displacement with respect to time, and vibration acceleration sensors, the rate of change of velocity. These three parameters that characterize vibration are interrelated, and by controlling, for example, vibration acceleration through single or double integration, it is easy to calculate the remaining two parameters.

The presence of three types of sensors is required to control vibration at facilities with different frequency characteristics. Vibration displacement sensors are well-proven in the low-frequency region, vibration velocity sensors are typically used for medium-frequency facilities, and vibration acceleration sensors are used for high-frequency ones.

Based on the results of our analysis of the possible applications of vibration sensors to monitor the beginning of changes in the technical condition of drilling rigs, BeanDevice AX-3D sensors were found most suitable. These sensors can be easily installed in the most vulnerable spots of the rig structure. The important factor in choosing this sensor was the possibility to collect measurement information from the sensors via Wi-Fi by means of a BeanGateway controller. The range of Wi-Fi signals from the BeanDevice AX-3D sensor is up to 650 m, which is sufficient, considering the size of a drilling rig. The technical parameters of the BeanDevice AX-3D sensor are described in [21].

In conclusion, we shall point out the following specifics of the problem of ensuring the safety of the drilling process.

1. Oil and gas extracting drilling rigs operate under continuous oscillating conditions due to changes in soil composition and other factors at different depths, and the most information about the beginning of the latent period of an emergency state of drilling rigs is contained in vibration signals.
2. The use of traditional algorithms and technologies for spectral analysis of noisy signals in drilling rig control and diagnostics systems is effective and expedient in the absence of correlation between the useful signal and the noise. They have found wide practical application, as this condition is met for many signals received at the outputs of corresponding sensors in the process of drilling. However, in certain cases, a correlation appears between the useful signal and the noise in the initial period of transition of the drilling rig into an emergency state. If delayed indication of malfunction does not result in an emergency state of the facility (if the driller manages to take appropriate measures to prevent an accident), then the use of traditional algorithms can also be considered expedient. However, in cases when, due to the delayed indication of the onset of malfunctions, the driller does not manage to prevent catastrophic accidents, in addition to the traditional algorithms for monitoring and diagnosing the technical condition of drilling rigs, it is advisable to use the proposed algorithms and noise technologies.
3. At the beginning of the latent period of an emergency condition of drilling rigs, noises correlated with the useful vibration signals generate, and these noises in some cases re the only source of diagnostic information about the beginning of the latent period of the facility's transition to an emergency state. However, in existing drilling rig control and management systems, as a result of filtration, this valuable information is lost. The use of the technology for forming the informative attributes from the noise estimates in combination with traditional algorithms will help ensure the normal operation of drilling rigs. The use of the noise control subsystem (Fig. 5.5) as a toolkit can free the driller from stressful and exhausting work that requires constant attention associated with the specifics of sudden occurrence of dangerous accident situations. At the same time, the driller is responsible for making decisions to take measures to eliminate the causes of the presumed accident. However, in this case, it becomes possible to reduce the dependence of safety of the drilling process on the drillers qualification.

References

1. Aliev TA, Alizada TA, Rzayeva NE et al (2017) Noise technologies and systems for monitoring the beginning of the latent period of accidents on fixed platforms. Mech Syst Signal Process 87:111–123. https://doi.org/10.1016/j.ymssp.2016.10.014
2. Wikipedia: Oil platform. https://en.wikipedia.org/wiki/Oil_platform. Accessed 21 May 2018
3. Oldest offshore oil platform. www.guinnessworldrecords.com/world-records/oldest-offshore-oil-platform-/. Accessed 21 May 2018
4. Types of offshore oil and gas structures. www.oceanexplorer.noaa.gov/explorations/06mexico/background/oil/media/types_600.html. Accessed 21 May 2018
5. Hibernia Construction. www.hibernia.ca/exploration.html. Accessed 21 May 2018
6. SOCAR prepares schedule to rectify accident at Bulla-Deniz offshore gas field. http://en.trend.az/business/energy/2180565.html. Accessed 21 May 2018
7. Gas Flow Resumes from Bulla-Deniz. Fire Extinguished (Azerbaijan). www.offshoreenergytoday.com/gas-flow-resumes-from-bulla-deniz-fire-extinguished-azerbaijan/. Accessed 21 May 2018
8. Fire extinguished on two gas wells of Azerbaijan's Guneshli platform. https://en.azvision.az/news.php?id=26110. Accessed 21 May 2018
9. Current situation in 'Guneshli' field. https://en.azvision.az/news.php?id=24849. Accessed 21 May 2018
10. Azerbaijani platform head discloses incident details. https://en.azvision.az/news.php?id=24816. Accessed 21 May 2018
11. Gunashli Platform No.10 fire. https://en.wikipedia.org/wiki/Gunashli_Platform_No.10_fire. Accessed 21 May 2018
12. May P, Mendy G, Tallett P (2009) Structural integrity monitoring: review and appraisal of current technologies for offshore applications. Prepared by Atkins Limited for the Health and Safety Executive. www.hse.gov.uk/research/rrpdf/rr685.pdf. Accessed 21 May 2018
13. FUGRO Offshore Structural Monitoring. www.fugro.com/our-services/marine-asset-integrity/monitoring-and-forecasting/offshore-structural-monitoring. Accessed 21 May 2018
14. BMT's Offshore Platform Current Monitoring. www.scimar.com/media/576648/Offshore%20Platform%20Current%20Monitoring.pdf. Accessed 21 May 2018
15. BMT's Independent Remote Monitoring. www.scimar.com/media/576640/Independent%20Remote%20Monitoring%20System.pdf. Accessed 21 May 2018
16. BMT's Subsea Strain Sensor Assembly. www.scimar.com/media/576664/Subsea%20Strain%20Sensor%20Assembly_2010.pdf. Accessed 21 May 2018
17. Aliev TA, Mamedov SI (2002) Telemetric information system to prognose accident when drilling wells by robust method. Oil Ind J 3
18. Dong GJ, Chen P (2016) A review of the evaluation, control, and application technologies for drill string vibrations and shocks in oil and gas well. Shock Vib 2016:1–34. https://doi.org/10.1155/2016/7418635
19. Dong G, Chen P (2018) The vibration characteristics of drillstring with positive displacement motor in compound drilling. Part 1: Dynamical modelling and monitoring validation. Int J Hydrogen Energy 43(5):2890–2902. https://doi.org/10.1016/j.ijhydene.2017.12.161
20. Ghasemloonia A, Rideout DG, Butt SD (2015) A review of drillstring vibration modeling and suppression methods. J Petrol Sci Eng 131:150–164. https://doi.org/10.1016/j.petrol.2015.04.030
21. Wireless accelerometer BeanDevice AX-3D. www.bestech.com.au/wp-content/uploads/BeanDevice-AX-3D.pdf. Accessed 21 May 2018

Chapter 6
The Use of Noise Control Technology and System at Oil and Gas Production Facilities

Abstract It is shown that the known systems of control and diagnostics of sucker rod pumping units of oil wells do not always succeed in early detection of the initiation of defects, at which negative consequences do not manifest themselves. This is due to the fact that in these systems, information about this is generated only at the beginning of a pronounced malfunction. A technology is proposed for forming the set of combinations of reference informative attributes from the estimates of noise characteristics and the normalized correlation functions of the load signal received from load cell mounted on the hanger of a sucker rod pumping unit. It is shown that each of them is indicative of one of the possible technical conditions and can therefore be used to solve problems of control, identification, and management of the oil production process. The possibility of controlling the start of the vibration state of compressor stations based on the estimates of the noise variance and the cross-correlation function between the noise and the useful signal obtained from the vibration sensors installed in the most informative elements of a station is also shown. Some results of the integration of the proposed systems into real-life facilities are given, showing that both the system for identifying the beginning of the latent period of changes in the technical condition of sucker rod pumping units and the system for monitoring the beginning of changes in the vibration state of compressor stations can find wide industrial application.

6.1 Noise Control System of Sucker Rod Pumping Unit

The use of sucker rod pumping units (SRPU) is known to be the primary method in artificial lift. Sucker rod pumping is widespread in the world oil production practice at present, covering over 85% of the total active wells stock in the USA [1, 2]. The popularity of the method is due to its simplicity, reliability, and applicability in a wide variety of operating conditions.

However, every day, with the decreasing oil reserves, increased reservoir flooding and well shutdowns caused by the inadequate identification of the technical condition of equipment, the profitability of oil production by SRPU decreases considerably. Therefore, improved adequacy of identification of the technical condition of SRPU is the main issue in ensuring profitability of oilfields in long-term operation. By resolving this issue, we can manage SRPU in real time, which can ensure the necessary stabilization of oil production. To increase the overhaul period and create the most favorable conditions for oil production management, various methods and tools of control of technical condition and management of SRPU have been proposed over the last several decades [3–9]. The results of these studies showed that the load in the rod suspension point contains the comprehensive and least distorted data on the condition of the underground pumping equipment. Therefore, dynamometry, i.e., reading and analysis of the curve of the load $U_p(t)$ received from the load cell in the rod suspension point $P(S)$ is considered the common way to control the technical condition of SRPU.

The authors of [3] demonstrate the possibility of recognition of load curves $U_p(t)$ with the use of low-frequency spectrum analyzers. The possibility of obtaining the amplitude spectrum of the dynamometer card is considered an advantage of this method. For instance, it was revealed that dynamometer cards of pump's normal operation have no even harmonics, while dynamometer cards of leaky pumps have even harmonics, whose amplitudes depend heavily on the scale of the leakage. However, the number of recognizable fault types was only four [3]. The statistical method was also used for fault recognition by dynamometer cards. This method compares favorably with other methods of dividing dynamometer cards into classes due to small amounts of computation and memory its application requires. The authors of [3, 5–7] give detailed description of the results of numerous studies, which have been carried out in this field over many years. Some or other abovementioned identification methods have been used at different SRPU control stations at real oilfields for a long time. The scientific foundations have been formed on the basis of these works and various systems for SRPU control and management by means of dynamometer cards obtained at the wellhead.

On the basis on the results of operation of those systems, dynamometer card based identification methods have been categorized as follows [3]:

- identification based directly on the characteristics of the ground dynamometer card;
- identification based on the secondary characteristics of the ground dynamometer card (spectral characteristics: variance, correlation, and regression of the signal of the load cell, coefficients of Fourier series expansion for the dynamometer card, etc.);
- identification based on the typical characteristics of the shape of the ground dynamometer card;
- identification by comparison of the shape of the dynamometer card under investigation with the reference one taken immediately after the repair of the well and stored in the device memory;
- identification based on the characteristics of the plunger dynamometer card calculated from the data of the ground dynamometer card and well design;
- identification based on the typical characteristics of the shape of the plunger dynamometer card.

The shortcoming of all these methods is that they do not allow performing automatic identification of a dynamometer card in real-time mode with sufficient adequacy. For this reason, the identification of dynamometer cards in real life is mostly performed by interpreting it in the semiautomated mode, which eventually comes down to the visual analysis of the obtained dynamometer information by a technologist, who makes the final decision on the presence of a fault in SRPU. The results depend on the qualification of the technologist and the diagnostics of all wells takes rather a long time. Besides, even a highly qualified specialist sometimes cannot determine precisely the technical condition of a deep well pump visually only from dynamometer cards, particularly for deep wells. Therefore, new technologies for real-time analysis and identification of dynamometer cards with the use of modern controllers have to be developed. In this case, it is appropriate to ensure the monitoring of changes in the technical condition of SRPU by identifying the signal of the load on the rod hanger per pumping cycle. Our research has demonstrated that one of the most efficient ways to solve this problem is to use a technology for identifying load signals combined with the methods of correlation analysis [6–8].

In the known SRPU control systems, the dynamometer card information comes from load cells and stroke sensors in the form of the electric signal of load $U_p(t)$ and stroke $U_s(t)$. via the communication channel. Using the combinations of these two variables $U_p(t)$ and $U_s(t)$, the dynamometer card $U_p(t) = f(U_s)$, whose form is described by a parallelogram (Fig. 6.1), is formed. Technologist determines more than 20 types of the technical condition of SRPU by visual analysis of distortions in different sections of its shape [3]. However, performing this operation for a hundred wells in real time requires automated identification.

It should be noted that in case of hardware implementation of identification of facility's state, there is no need to use $U_s(t)$, since this can be accomplished by analyzing only the stress curve $U_p(t)$. The main challenge of this problem is currently associated with the lack of a technology that allows ensuring the adequacy of iden-

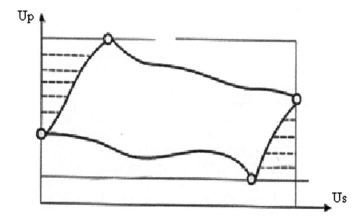

Fig. 6.1 Typical dynamometer card of normal operation of SRPU

tification of the load signal $U_p(t)$ in real time. For instance, when the correlation analysis technology is used for this purpose, the condition of robustness is not fulfilled, because the error of the obtained estimates of correlation functions caused by the effects of the noise $\varepsilon_1(t)$ accompanying the useful load signal $U_p(t)$ under operation changes within quite a wide range. This is because the control object, i.e., SRPU, operates in the field environment (temperature and humidity extremes, winds, etc.). Besides, during the pump operation, various faults also cause the formation of the noise $\varepsilon_2(t)$ that correlates with the signal $U_p(t)$ [10]. Therefore, the noise $\varepsilon(t)$ accompanying the load signal $U_p(t)$ forms under the influence of the following two factors:

$$\varepsilon(t) = \varepsilon_1(t) + \varepsilon_2(t),$$

where $\varepsilon_1(t)$ forms due to the changes in the environment (temperature and humidity differences, etc.); $\varepsilon_2(t)$ forms during the operation of the object due to the changes in the technical condition of the mechanical components of the pump, such as wear and tear, bends, cracks, fatigue, etc.

Thus, during the operation of objects, the signal contaminated with the noise $\varepsilon(t)$ comes to the input of the system instead of the signal $U_p(t)$. The analyzed signal in the analog form is as follows:

$$g(t) = U_p(t) + \varepsilon(t),$$

and in the digital form is as follows:

$$g(i\Delta t) = U_p(i\Delta t) + \varepsilon(i\Delta t),$$

where i is the serial number of the measurement of the signal's analog-to-digital conversion; Δt is the sampling interval of the signal at the analog-to-digital conversion.

For said reasons, both the amplitude and the spectrum of the noise $\varepsilon(i\Delta t)$ vary in quite a wide range. The errors of the obtained estimates of the correlation functions $R_{gg}(i\Delta t)$ of the measuring signal $g(i\Delta t)$ also vary in time in a wide range due to the abovementioned reasons. Therefore, we fail to ensure the condition of robustness for the estimates of the correlation function in real time, i.e., to eliminate the dependence of the obtained results on the effects of the noise $\varepsilon(i\Delta t)$. This, in turn, complicates solving the problem of identification of the dynamometer card with the use of correlation methods. Consequently, ensuring the adequacy of identification requires that the conditions of robustness hold, i.e., that the effects of said factors on the error of the estimates $R_{gg}(i\Delta t)$ be eliminated.

At first glance, the effects of the errors on the results of identification of dynamometer cards can be eliminated by filtration of the noise accompanying the useful signal $U(i\Delta t)$. If the noise spectrum is stable, the use of filtration usually produces satisfactory results. However, in the field conditions, the spectrum of the noise changes in a wide range due to the abrupt change of the factors of its formation.

6.1 Noise Control System of Sucker Rod Pumping Unit

Besides, the variance of the spectrum of the noise that forms due to mechanical processes in SRPU changes in a wide range as well, frequently overlapping the range of the useful signal spectrum. For these reasons, we cannot achieve the desired effect by using the technology for filtering the load signal. Because of this, we cannot always achieve satisfactory results through correlation analysis of the force signal with the use of filtration. Therefore, solving the problem under consideration first of all requires developing the technologies for calculating such estimates of correlation characteristics that practically cannot be affected by changes in said noises.

To that end, it is appropriate first to reduce the estimates $R_{gg}(i\Delta t)$ to a single dimensionless value by applying the normalization procedure [6, 10–13]. However, our analysis demonstrates that the application of traditional methods introduces additional error in the normalized estimates of the correlation functions $r_{gg}(i\Delta t)$, which, in turn, complicates ensuring adequacy of the results of solving the abovementioned problems. Let us consider this issue in greater detail.

It is known that the normalized correlation functions of the dynamometer card of $U(i\Delta t)$ are calculated from the following formula [10–12]:

$$r_{UU}(\mu) = R_{UU}(\mu)/D_U = R_{UU}(\mu)/R_{UU}(\mu = 0), \qquad (6.1)$$

where the estimate of the variance $D_U = R_{UU}(\mu)$ at $\mu = 0$ is calculated from the expression

$$R_{UU}(\mu = 0) = D_U = \frac{1}{N}\sum_{i=1}^{N} U(i\Delta t)U(i\Delta t), \quad \mu = 0, 1, 2, 3, \ldots, N,$$

where N is the number of measurements of analog-to-digital conversions of the signal per pumping cycle of SRPU.

The estimates of the correlation function $R_{UU}(\mu)$ of the useful signal $U_p(i\Delta t)$ are calculated from the formula

$$R_{UU}(\mu) = \frac{1}{N}\sum_{i=1}^{N} U(i\Delta t)U((i+\mu)\Delta t), \quad \mu = 0, 1, 2, 3.$$

It is also known that the estimates of the normalized correlation functions $r_{gg}(\mu)$ of the noisy dynamometer card $g(i\Delta t)$ are calculated from the formula

$$r_{gg}(\mu) = \frac{R_{gg}(\mu)}{D_g} = \frac{R_{gg}(\mu)}{R_{gg}(\mu = 0)}. \qquad (6.2)$$

To ensure the adequacy of the results of identifying the technical condition of SRPU, it is necessary that the normalization of the correlation functions of the signal of the dynamometer card $g(i\Delta t)$ from expressions (6.1) and (6.2) resulted in the fulfillment of the condition

$$r_{UU}(\mu) \approx r_{gg}(\mu).$$

It is obvious that when $\mu = 0$, the normalization results obtained from expressions (6.1) and (6.2) will match, i.e.,

$$r_{UU}(\mu = 0) = \frac{R_{UU}(\mu = 0)}{D_U = r_{gg}(\mu = 0)} = \frac{R_{gg}(\mu = 0)}{D_g} = 1$$

However, it is also obvious that the normalization results for $\mu \neq 0$ will differ, i.e.,

$$r_{UU}(\mu) = R_{UU}(\mu)/R_{UU}(\mu = 0) \neq r_{gg}(\mu) = R_{gg}(\mu)/R_{gg}(\mu = 0).$$

Therefore, when formula (6.2) is used, the correct result is obtained only at $\mu = 0$. For all other cases, when $\mu \neq 0$, the results of the normalization of the correlation functions of the noisy signal differ from the results of the normalization of the correlation functions of the useful signal. This is the main reason for the appearance of an error in the normalization of the dynamometer card. Obviously, in order to eliminate this error by the traditional normalization procedure, it is appropriate to reduce the expression (6.2) to the following form:

$$r_{gg}(\mu \neq 0) = \frac{R_{gg}(\mu \neq 0)}{R_{UU}(\mu = 0)}. \tag{6.3}$$

However, as was pointed out earlier, it is practically impossible to calculate the estimate $R_{UU}(\mu = 0)$ due to the effects of the noise $\varepsilon(i\Delta t)$.

In view of the above, the creation of a SRPU management system requires developing a technology for eliminating the effects of the noise $\varepsilon(i\Delta t)$ in the estimates of the normalized correlation functions $r_{gg}(\mu \neq 0)$ of real dynamometer cards. We also have to obtain such estimates of the normalized correlation functions $r_{gg}(\mu)$ of the signals $g(i\Delta t)$ that would ensure that the following equality holds:

$$r_{gg}(\mu) \approx r_{UU}(\mu).$$

Taking into account that

$$R_{gg}(\mu \neq 0) \approx R_{UU}(\mu \neq 0)$$

Formula (6.3) for estimating the normalized correlation function of the dynamometer card $g(i\Delta t)$ can be written as

$$r^R_{UU}(\mu \neq 0) = \frac{R_{gg}(\mu \neq 0)}{D_U} = \frac{R_{UU}(\mu \neq 0)}{D_U}.$$

Considering that the variance D_g of the noisy signal is [10, 13]

6.1 Noise Control System of Sucker Rod Pumping Unit

$$D_g = D_U + D_\varepsilon,$$

the formula for calculating the normalized correlation function can be written as

$$r_{UU}(\mu \neq 0) = \frac{R_{gg}(\mu \neq 0)}{D_U - D_\varepsilon}.$$

This makes it possible, by eliminating the additional error caused by the effects of the noise, to ensure sufficient accuracy of estimates of the normalized correlation functions $r_{UU}(\mu \neq 0)$ from the formula

$$r_{UU}(\mu) = \begin{cases} \frac{R_{UU}(\mu=0)}{D_g} = 1 & \text{when } \mu = 0 \\ \frac{R_{UU}(\mu)}{D_U - D_\varepsilon} & \text{when } \mu \neq 0 \end{cases} \quad (6.4)$$

where $R_{UU}(\mu)$ is the estimates of the normalized correlation functions $r_{gg}(\mu)$ of the dynamometer card at $\mu = 0, 1, 2, 3, \ldots$.

Our experiments in real oilfields have shown that the proposed procedure allows for calculating the estimates (free from the noise-induced errors) of the normalized autocorrelation function $r_{gg}(\mu)$ of the dynamometer card, i.e., of the load signal $g(i\Delta t)$ obtained at the output of the load cell of the rod string on the hanger of SRPU.

Our test application of this technology and the control system have established that it is possible to form up to 11 informative attributes, using the normalized correlation functions $r_{UU}(\mu)$ of the dynamometer card $g(i\Delta t)$, which are calculated from expression (6.4). In order to identify the technical condition of SRPU, it is appropriate to use the procedure for calculating the informative attribute of the dynamometer card. To this end, first, the pumping cycle T_{ST} of SRPU is divided into 8 equal time intervals ΔT_{ST} that are calculated from the formula

$$\Delta T_{ST} = \frac{T_{ST}}{8}$$

For instance, if $T_{ST} = 1024$ samples of the discrete values of the load signal are taken per pumping cycle of SRPU, then ΔT_{ST} will be equal to 128 samples.

Further, using Formulas (6.4), the estimates $r_{UU}(\mu = 0)$, $r_{UU}(\mu = 1\Delta T_{ST})$, $r_{UU}(\mu = 2\Delta T_{ST}), \ldots, r_{UU}(\mu = 7\Delta T_{ST})$ are calculated, and their differences are calculated from the expressions

$$\Delta r_{UU}(\mu = 1\Delta T_{ST}) = r_{UU}(\mu = 0) - r_{UU}(\mu = 1\Delta T_{ST}),$$
$$\Delta r_{UU}(\mu = 3\Delta T_{ST}) = r_{UU}(\mu = 2\Delta T_{ST}) - r_{UU}(\mu = 3\Delta T_{ST}),$$
$$\Delta r_{UU}(\mu = 5\Delta T_{ST}) = r_{UU}(\mu = 4\Delta T_{ST}) - r_{UU}(\mu = 5\Delta T_{ST}),$$
$$\Delta r_{UU}(\mu = 7\Delta T_{ST}) = r_{UU}(\mu = 6\Delta T_{ST}) - r_{UU}(\mu = 7\Delta T_{ST}).$$

The next step is to calculate the minimum value of the normalized correlation function of the load signal $r_{UU\min}(\mu)$ and the value $\mu_{\min}$ corresponding to it, after which the informative attributes are determined in the form of the following coefficients:

$$K_{N1} = \Delta r_{UU}(\mu = 1\Delta T_{ST}); \quad K_{N2} = \Delta r_{UU}(\mu = 3\Delta T_{ST});$$
$$K_{N3} = \Delta r_{UU}(\mu = 5\Delta T_{ST}); \quad K_{N4} = \Delta r_{UU}(\mu = 7\Delta T_{ST});$$
$$K_{N5} = \frac{\Delta r_{UU}(\mu = 1\Delta T_{ST})}{\Delta r_{UU}(\mu = 3\Delta T_{ST})}; \quad K_{N6} = \frac{\Delta r_{UU}(\mu = 1\Delta T_{ST})}{\Delta r_{UU}(\mu = 5\Delta T_{ST})};$$
$$K_{N7} = \frac{\Delta r_{UU}(\mu = 1\Delta T_{ST})}{\Delta r_{UU}(\mu = 7\Delta T_{ST})}; \quad K_{N8} = \frac{\Delta r_{UU}(\mu = 3\Delta T_{ST})}{\Delta r_{UU}(\mu = 5\Delta T_{ST})};$$
$$K_{N9} = \frac{\Delta r_{UU}(\mu = 3\Delta T_{ST})}{\Delta r_{UU}(\mu = 7\Delta T_{ST})}; \quad K_{N10} = \frac{\Delta r_{UU}(\mu = 5\Delta T_{ST})}{\Delta r_{UU}(\mu = 7\Delta T_{ST})};$$
$$K_{N11} = \Delta r_{UU\min}(\mu).$$

Further, we will demonstrate the possibility of using the coefficients $K_{N1} - K_{N11}$ as informative attributes for identifying the technical condition of SPRU.

They have been used in the complex of control, diagnostics, and management for oil wells operated by sucker rod pumping units at Bibi-Heybat Oil and Gas Production Department [1, 14].

The simplicity of calculation of those coefficients makes it possible to implement the proposed identification technology by means of inexpensive modern industrial controllers (in our case, LPC 2148 FBD64 controller was used).

Figures 6.2 and 6.3 show the block diagram of the SRPU management system and the external appearance of the robust management station (RMS) on the beam-pumping unit. The diagram consists of three levels:

1. The level of the deep well pumping unit consisting of plunger pump 1; plunger 2; tubing 3; rods 4; polished rod 5; horsehead 6; walking beam 7; pitman 8; crank counterweight 9; reductor 10; multiple V-belt drive 11; electric motor 12; beam equalizer 13; load cell 14; wellhead pressure sensor 15; rotation angle sensor 16; crank of the beam-pumping unit 17.
2. The level of RMS consisting of the controller for data acquisition from load cells 14; wellhead pressure sensor 15 and rotation angle sensor 16; frequency converter for controlling the speed of the electric motor; a wireless modem equipped with an antenna to provide data exchange via MODBUS-RTU protocol between RMS and the centralized control station.
3. The level of the centralized control station of the oil field, which serves up to 200 wells and consists of an industrial computer and a wireless modem with an antenna.

The SRPU management system employs the technology for identifying the dynamometer card based on the estimates of the normalized correlation functions of the load signal $U_p(i\Delta t)$. To implement them, the sampling interval of the signal was

6.1 Noise Control System of Sucker Rod Pumping Unit

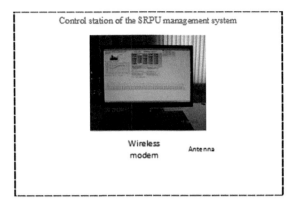

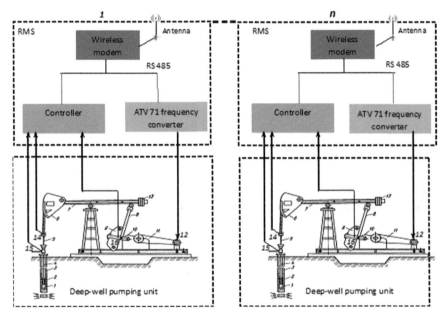

Fig. 6.2 Block diagram of the SRPU management system

determined first, on the basis of the duration of the pumping cycle T_{ST}. For most oil wells, the duration of T_{ST} varies within the range of 5–20 s. We have deduced from experiments that to obtain the estimate of the robust correlation function $r_{UU}^R(\mu)$ with the required accuracy, we only need to sample the load signal at the frequency $f = 500 - 100$ Hz. Our experiments have established that any minute change in the technical condition of SRPU during the pumping cycle affects the values of the estimates of the normalized correlation functions of the load signal $U_p(i\Delta t)$, which, in turn, leads to changes in the informative attributes $K_{N1} - K_{N11}$. As a result of SRPU operation, corresponding reference combinations of these coefficients are

Fig. 6.3 External appearance of RMS on the beam-pumping unit

formed and saved for various technical conditions of SRPU (Tables 6.1 and 6.2). Our experiments have shown that they allow for reliable identification of the load signal $U_p(i\Delta t)$, i.e., the technical condition of SRPU in real time. The identification of the technical condition of SRPU comes down to forming the combinations of current informative attributes and comparing them to the respective combinations of reference informative attributes $K_{N1} - K_{N11}$, which makes visual interpretation of the dynamometer card unnecessary for determining the current technical condition of SRPU. To illustrate the possibilities of the considered identification option in real industrial practice, we give 11 most common fault types in Fig. 6.4. Tables 6.1 and 6.2 contain combinations of the corresponding estimates of normalized coefficients.

In the field conditions of the system operation, the signals received from load cell 14 (Fig. 6.2) are used to calculate the combinations of the robust informative attributes $K_{N1} - K_{N11}$ in real time during the pumping cycle of SRPU. These coefficients are compared with the reference ones, which were determined in advance. For instance, they are given for one facility in Tables 6.1 and 6.2. The current technical condition of SRPU is determined on the basis of the coefficients that are closest to the reference coefficients. Thus, during the operation of SRPU, control commands for the facility are formed based on the results of identification, e.g., to change the duration of the pumping cycle. At the same time, the information on the state of the facility is transmitted to the control station via RMD 400 SP4 wireless modem (Fig. 6.2). When the object is in the normal state, this information is displayed on the screen at the control station of the management system in green color, when the initial stage of a fault is detected, this information is displayed in yellow color, and an emergency state is indicated with red color.

6.1 Noise Control System of Sucker Rod Pumping Unit

Table 6.1 Combinations of the robust informative attributes $K_{N1}^{R} - K_{N6}^{R}$

No	Fault type	Attribute					
		K_{N1}	K_{N2}	K_{N3}	K_{N4}	K_{N5}	K_{N6}
1	Breakdown of the fluid end of the pump	0.404963	0.529615	−0.327657	−0.568131	0.764637	−1.235937
2	Breakdown of the receiving end of the pump	0.475903	0.377308	−0.231654	−0.555379	1.261314	−2.054368
3	Plunger sticking	0.306373	0.642464	−0.274895	−0.661106	0.476871	−1.114508
4	Pump down of the level	0.501178	0.354056	−0.119199	−0.601160	1.415534	−4.204551
5	High plunger fit	0.400730	0.557043	−0.338440	−0.574514	0.719389	−1.184050
6	High percentage of leakage in the fluid end of the pump	0.375797	0.537027	−0.175253	−0.657772	0.699772	−2.144306
7	Leakage in the receiving end of the pump	0.418711	0.600065	−0.228021	−0.553510	0.697776	−1.836283
8	High percentage of leakage	0.427557	0.508795	−0.190109	−0.595119	0.840332	−2.249014
9	Normal operation of SRPU	0.447212	0.5587327	−0.417786	−0.548613	0.800413	−1.070433
10	Pumping off of gas and leakage in the fluid end of the pump	0.351550	0.623045	−0.293039	−0.669873	0.564244	−1.199669

Table 6.2 Combinations of the robust informative attributes $K_{N7}^R - K_{N11}^R$

№	Fault type	Attribute				
		K_{N7}	K_{N8}	K_{N9}	K_{N10}	K_{N11}
1	Breakdown of the fluid end of the pump	−0.712799	−1.616371	−0.932206	0.576728	−0.829877
2	Breakdown of the receiving end of the pump	−0.856899	−1.628753	−0.679370	0.417110	−0.639319
3	Plunger sticking	−0.463424	−2.337126	−0.971802	0.415811	−0.883965
4	Pump down of the level	−0.833685	−2.970294	−0.588954	0.198282	−0.574909
5	High plunger fit	−0.697512	−1.645911	−0.969590	0.589090	−0.870091
6	High percentage of leakage in the fluid end of the pump	−0.571318	−3.064291	−0.816434	0.266435	−0.745240
7	Leakage in the receiving end of the pump	−0.756466	−2.631622	−1.084110	0.411955	−0.799754
8	High percentage of leakage	−0.718439	−2.676340	−0.854947	0.319446	−0.720893
9	Normal operation of SRPU	−0.815170	−1.337350	−1.018436	0.761533	−0.972479
10	Pumping off of gas and leakage in the fluid end of the pump	−0.524801	−2.126153	−0.930095	0.437454	−0.937605

6.1 Noise Control System of Sucker Rod Pumping Unit

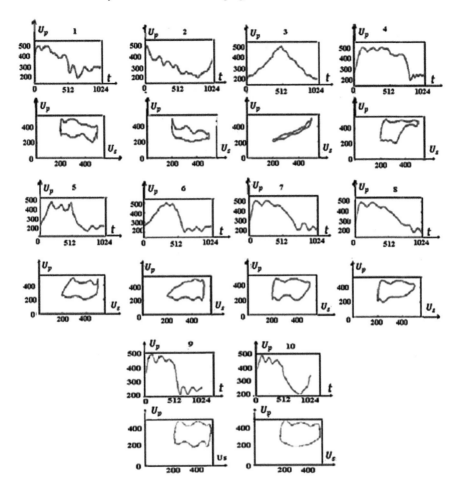

Fig. 6.4 Reference load curves and dynamometer cards

When the proposed technologies are introduced into oilfields with a large number of wells, combinations of corresponding reference coefficients for various technical conditions of SRPU are calculated and saved for each of them successively in the process of operation. This is implemented with the participation of a technologist, who identifies the technical condition of SRPU by interpreting the dynamometer card and registers its correspondence to the respective combinations of the robust informative attributes $K_{N1} - K_{N11}$ in the identification block of the system. Thus, in the process of SRPU operation, as various technical conditions arise at each well, a combination of corresponding reference estimates of coefficients is formed in the identification block of each system. As a result, after a certain period of operation, combinations of reference coefficients of corresponding technical conditions are formed and saved in the identification units of the SRPU robust management systems at all wells. After a certain period of operation, when the references of all possible conditions of SRPU have been formed and saved in the identification blocks, the system goes into the automated identification and management mode.

Figure 6.4 shows the reference load curves $U_p(i\Delta t)$ and corresponding dynamometer cards for ten typical technical conditions of SRPU. Given in Tables 6.1 and 6.2 are the combinations of the robust informative attributes $K_{N1} - K_{N11}$. It is obvious that only one combination of the estimates $K_{N1} - K_{N11}$ corresponds to each dynamometer card. It is also obvious that if the combinations of the robust informative attributes $K_{N1} - K_{N11}$ are available, we can definitely determine the same technical conditions that are usually determined by a technologist from a dynamometer card visually. Thus, the combinations of these coefficients make it possible to perform automated identification of the technical condition of SRPU. Consequently, it is no longer necessary to use dynamometer cards for visual identification of the technical condition of SRPU in the semiautomated mode.

Thus, it has been established experimentally that it is possible to determine the normalized coefficients $K_{N1}, K_{N2}, \ldots, K_{11}$, which are practically unaffected by the abovementioned noises, by means of normalized correlation functions. The advantage of using these coefficients is that they are easy to calculate on modern controllers (e.g., LPC 2148 FBD64). Thus, it becomes possible to perform diagnostics in real time. The simplicity of implementation of these technologies allows developing a simple, reliable, and inexpensive SRPU management system, which we have introduced at real facilities in Azerbaijan. The service experience of the system at 35 wells of Bibi-Heybat oilfield, 190 wells of the Azerbaijan-British company Shirvan Oil, etc., has demonstrated the reliability of these systems. The improved diagnostics and management of SRPU made it possible to operate the wells in an adequate mode and increase their profitability due to energy saving and prolonged overhaul period. For instance, energy saving reached 50% and overhaul period increased by 30% at Bibi-Heybat oilfield.

It should be noted that it has been established on the basis of the operational experience of these systems that the values of the same coefficients $K_{N1} - K_{N11}$ for the same technical conditions vary insignificantly (no more than 5–10%) for wells of the same depth at all oilfields. Therefore, if we determine combinations of these reference coefficients for corresponding technical conditions for one well, they can be used in SRPU management systems of other wells of similar depth. Considering that the pump is placed approximately at the same depth at most old deposits, it becomes obvious that the formation of the reference base of combinations of coefficients in SRPU management systems will not take much time.

6.2 System for Noise Control of the Beginning of the Latent Period of Accidents at Compressor Stations

At present, compressor stations are widely used in oil fields, oil and gas pipelines, and oil refineries. During the operation of compressor stations, after the period T_0 of normal functioning in seismic regions, the period T_1 of the latent transition into an emergency state begins. Then they go into the time period T_2 of the pronounced emergency state. Naturally, the periods T_0 and T_1 for different facilities can last months, days and even hours, depending on the technical condition and operating conditions. The time T_2 is different for different compressors as well. Despite all these differences, the control problem in all these cases comes down to ensuring a reliable indication of the beginning of the time T_1 of the latent period of compressor's transition into an emergency state begins. In control systems today, the registration of the beginning of the latent period T_1 of facility's emergency state is not ensured. This is one of the serious shortcomings of modern control systems. Shown below is one of the variants of experimental application of the noise control technology at real-life industrial facilities. For instance, the system for control of the beginning of transition into an emergency state at the compressor unit in Fig. 6.5 has been in operation at an oil refinery since 25 May 2007.

As we can see in Fig. 6.5, MK 301/2 compressor consists of an electric motor (EM), reducers (R1, R2), a low-pressure cylinder (LPC) and a high-pressure cylinder (HPC). ASA-062 (Acceleration Sensors type ASA-062) vibration sensors by Bruel & Kjaer are located as shown in Fig. 6.5a.

Figure 6.6 shows the scheme of the control system of the MK 301/2 compressor of the fluid catalytic cracking unit at the Baku oil refinery. Vulnerable informative spots (points) of the MK 301/2 compressor unit are equipped with the following vibration sensors:

- on the electric motor (EM)—4 points (sensors 11, 12, 13, 14);
- on the first reducer (R1)—2 points (sensors 4,5);
- on the low-pressure cylinder (LPC)—2 points (sensors 6, 7);
- on the second reducer (R2)—2 points (sensors 9,10);
- on the high-pressure cylinder (HPC)—2 points (sensors 1, 2);
- on the supports—3 points (sensors 3, 8, 15).

The outputs of sensors 1–15 produce the vibration acceleration signals $g(i\Delta t)$. They arrive at the inputs of the controllers (Fastwel CPU188-5v.3) through the safety barriers (Safety Barriers AC-297, ATEX for ASA-062) and amplifiers (Conditioning Amplifier 2694-A) and are analyzed on the computer by means of the proposed technology.

At the time a defect emerges, the estimates of the noise characteristics $D_\varepsilon, R_{X\varepsilon}(\Delta t), R_{X\varepsilon}(2\Delta t), R_{X\varepsilon}(3\Delta t), \ldots$ of the signals $g_1(i\Delta t), g_2(i\Delta t), g_3(i\Delta t), \ldots$ at the outputs of the corresponding sensors differ from zero. For instance, in Fig. 6.7, the onset of defects from 16 May to 21 May and from 25 May to 26 May 2007 was registered as a result of the calculation of the estimate of the noise variance D_ε of

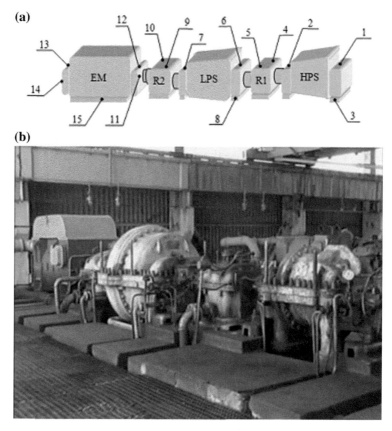

Fig. 6.5 MK 301/2 compressor: **a** the layout of sensors on the compressor; **b** the photo of the compressor of the fluid catalytic cracking unit at the Baku oil refinery

the signal coming from the first high-pressure cylinder sensor (HPC), which turned out to be different from zero. So did the estimates of the cross-correlation function between the useful signal and the noise $R_{X\varepsilon}(\Delta t)$ of the same signal. The analyzed vibration signals are sampled with the frequency 20 kHz, i.e., at the sampling interval $\Delta t = 50\,\mu S$. In every cycle, the observation time T is 0.5 s and $N = 10{,}000$ samples are processed over a period. These cycles are repeated continuously.

Figure 6.8 shows the graph of the vibration velocity in the specified period at the specified measurement point. It demonstrates the stability of the vibration state in contrast to Fig. 6.8. This is the advantage of the proposed technologies.

In conclusion, it should be noted that in their training period, these subsystems can calculate the combinations of the maximum threshold values of the estimates of all possible noise characteristics for all analyzed signals in the process of noise control of the facility in its normal operation mode. As a result, after a certain period of time, the reference set corresponding to facility's normal technical condition is

6.2 System for Noise Control of the Beginning of the Latent Period … 117

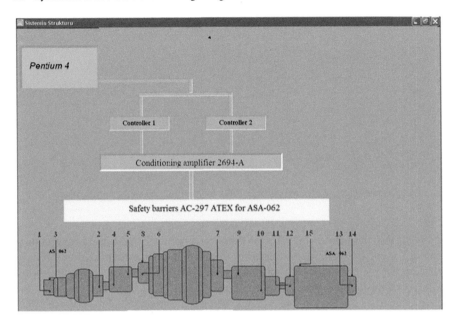

Fig. 6.6 Structure of the control system of the compressor unit

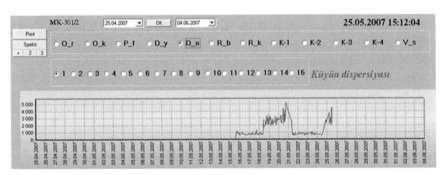

Fig. 6.7 Graph of the noise variance showing the initiation of a defect and a change in the vibration state

formed (and saved to memory) from these combinations of all possible maximum estimates, in the form

$$W_j^{\max} = \begin{cases} D_{g_j\varepsilon}^{\max}, R_{g_jX\varepsilon}^{\max}(1\Delta t), R_{g_jX\varepsilon}^{\max}(2\Delta t), R_{g_jX\varepsilon}^{\max}(3\Delta t), \ldots, R_{g_jX\varepsilon}^{\max}(m\Delta t) \\ R_{g_jXc}^{*\max}(1\Delta t), R_{g_jX\varepsilon}^{*\max}(2\Delta t), R_{g_jX\varepsilon}^{*\max}(3\Delta t), \ldots, R_{g_jX\varepsilon}^{*\max}(m\Delta t) \\ a_{g_jn}^{\max}, b_{g_jn}^{\max}; a_{g_jn\varepsilon}^{\max}, b_{g_jn\varepsilon}^{\max}; a_{g_jn\varepsilon_2}^{\max}, b_{g_jn\varepsilon_2}^{\max} \\ a_{g_jn}^{*\max}, b_{g_jn}^{*\max}; a_{g_jn\varepsilon}^{*\max}; b_{g_jn\varepsilon}^{*\max}; a_{g_jn\varepsilon_2}^{*\max}; b_{g_jn\varepsilon_2}^{*\max} \\ k_{g_jq_0}^{\max}, k_{g_jq_1}^{\max}, k_{g_jq_2}^{\max}, \ldots, k_{g_jq_m}^{\max}; \bar{f}_{g_jq_0}^{\max}, \bar{f}_{g_jq_1}^{\max}, \bar{f}_{g_jq_2}^{\max}, \ldots, \bar{f}_{g_jq_m}^{\max} \end{cases}$$

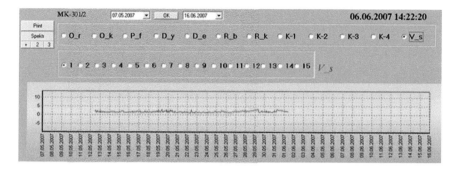

Fig. 6.8 Graph of the results of the vibration velocity analysis (shows the normal state)

When the subsystem is operating in the noise control mode during the operation of the equipment, the resulting current combination of the estimates of all possible noise characteristics of all analyzed signals is compared with the corresponding maximum reference threshold estimates, which are usually slightly different from zero. In this case, when the current estimates become equal to or greater than the corresponding reference threshold estimates, the result is interpreted as the beginning of the latent period of an emergency state.

During the operation of the intelligent noise control system shown in Fig. 6.5, similar cases have been recorded repeatedly over several years, which confirmed the effectiveness of the proposed technology. During this time, the described system never failed to signal and register a breach of facility's operating regulations when the compressor station (repeatedly) went into the period of the latent emergency state.

Due to this and by continuous control of the beginning of the latent period and the development dynamics of accidents of the compressor unit, it was possible to increase the level of safety and reliability of the facility's operation.

Thus, to indicate the start of the latent period of compressor's transition from the normal state into an emergency state and to determine the dynamics of its development, the control system uses the technology for calculating the estimates of the noise variance, the cross-correlation functions between the useful signal and the noise, which in the normal state take values equal to zero. In the process of facility's operation, a set of informative attributes are formed from them, and the number and magnitude of the nonzero element are used to determine the location, dynamics, and nature of the emerging fault. The experience of using proposed technologies at the compressor station since 2007 has shown their effectiveness. The system provides reliable signaling about the beginning of all accident situations. Due to this, the accident-free operation of the unit was ensured. This experience also demonstrates the possibility of using the noise control technologies at other similar technical facilities.

References

1. Aliev TA, Rzayev AH, Guluyev GA et al (2018) Robust technology and system for management of sucker rod pumping units in oil wells. Mech Syst Sig Process 99:47–56. https://doi.org/10.1016/j.ymssp.2017.06.010
2. Neely AB, Tolbert HE (1988) Experience with pump-off control in the Permian Basin, SPE Paper No. 14345. SPE technical conference and journal of petroleum technology, pp 645–648
3. Aliev TM et al (1988) Automated control and diagnostics of SRPU. Nedra, Moscow
4. Gibbs SG, Neely AB (1966) Computer diagnosis of down-hole conditions in sucker rod pumping wells. J Petrol Technol 8(01):91–94. https://doi.org/10.2118/1165-PA
5. Aliev TA, Nusratov OK (1998) Algorithms for the analysis of cyclic signals. Autom Control Comput Sci 32(2):59–64
6. Aliev TA, Guluyev GA, Pashayev FH et al (2012) Noise monitoring technology for objects in transition to the emergency state. Mech Syst Sig Process 27:755–762. https://doi.org/10.1016/j.ymssp.2011.09.005
7. Aliev TA, Abbasov AM, Guluyev QA et al (2013) System of robust noise monitoring of anomalous seismic processes. Soil Dyn Earthq Eng 53:11–25. https://doi.org/10.1016/j.soildyn.2012.12.013
8. Fasel TR, Todd MD (2010) Chaotic insonification for health monitoring of an adhesively bonded composite stiffened panel. Mech Syst Sig Process 24(5):1420–1430
9. Tang H, Liao YH, Cao JY, Xie H (2010) Fault diagnosis approach based on Volterra models. Mech Syst Sig Process 24(4):1099–1113
10. Aliev T (2007) Digital noise monitoring of defect origin. Springer, Boston. https://doi.org/10.1007/978-0-387-71754-8
11. Bendat JS, Piersol AG (2000) Random data, analysis & measurement procedures. Wiley, New York
12. Bendat JS, Piersol AG (1993) Engineering applications of correlation and spectral analysis, 2nd edn. Wiley, New York
13. Aliev T (2003) Robust technology with analysis of interference in signal processing. Kluwer Academic/Plenum Publishers, New York. https://doi.org/10.1007/978-1-4615-0093-3
14. Aliev TA, Iskenderov DA, Guluyev GA et al (2014) Results of introducing the control, diagnostics and management complex for oil wells operated by sucker rod pumps at "Bibi Heybat Oil" oil and gas producing company. Azerbaijan Oil Econ 6:37–41

Chapter 7
Possibilities of the Use of the Technology and System of Noise Control of the Beginning of the Latent Period of Accidents on Technical Facilities

Abstract The specifics of the emergence of accidents at power plants and the difficulties of their prediction are analyzed. It is shown that traditional technologies of analyzing the signals obtained from the sensors of these stations do not allow one to extract diagnostic information contained in the noise. Therefore, control systems cannot register the beginning of the latent period of changes in the technical condition of power plants. To eliminate this drawback, it is proposed to connect the subsystems of noise control of the beginning of the latent period of accidents to the existing system. This will improve the safety of operation of power plants. It is shown that the technologies and systems of noise control should also be used to create an intelligent system of preflight noise monitoring of aircraft's technical condition. The chapter also considers the possibility of using the technology for analyzing and extracting diagnostic information from the noise and the corresponding hardware in railway safety systems in seismically active regions by creating and using an intelligent noise control subsystem for the control of the technical condition of individual train cars and railway tracks, bridges and tunnels.

7.1 Specifics and Relevance of the Use of the Technology and System of Noise Control at Power Engineering Facilities

It is known that nuclear power engineering has the most potential for satisfying the society's ever-growing demand for energy. It is also a fact that this is only possible with guaranteed safety of operation of nuclear power plants. At the same time, ensuring safety of operation of hydroelectric and thermal power stations is also of enormous practical importance. Therefore, it is strategically important to develop fundamentally new information technologies and security control systems for nuclear power plants and hydroelectric and thermal power stations [1–4].

Various accidents at power stations are usually caused by various defects. The causes of their inception in various components of hydroelectric and thermal power stations and nuclear power plants are the subject of separate research. However, they have a lot in common in terms of acquisition of information on the beginning of initiation of these defect and methods of their analysis. In many cases, the outputs of sensors installed in appropriate components receive signals that reflect the beginning of defect initiation [1–5]. Monitoring and control systems of hydroelectric and thermal power stations and nuclear power plants, analyzing those signals as carriers of information, employ practically the same information technologies. These technologies are implemented on similar modern computers. Therefore, to an information measuring systems specialist, solving problems of control of the latent period of their emergency state by way of analysis of the signals received from sensors is not much different, despite the wide range of specific features of these power stations.

For these power stations, however, the process from origin and development of a defect to the moment it leads to a pronounced emergency state has significant distinctive features related to the physical, biological, mechanical, chemical, and other properties of power station equipment, as well as to the functions they fulfill, their operation modes, etc. For these and many other reasons, the process of defect origin before the defect takes an explicit form runs differently for different power engineering facilities. They have difference change dynamics. For some facilities, this is a fast process, for others, it takes much more time. Despite all the differences, the common fact is that the information contained in the noise of the signals received at the outputs of corresponding sensors changes continuously in that period due to the dynamics of defect origin and development. Thus, from the very beginning of the latent period of accidents to the moment it becomes pronounced, the noise often becomes the only carrier of valuable information. Unfortunately, the technologies used in the control of technological processes of power engineering facilities make no allowance for the aforementioned specifics of formation of real-life signals. Due to this, it is appropriate to use noise control of the beginning of the latent period of accidents in control and management systems of hydroelectric and thermal power stations and nuclear power plants often fail to ensure identification of changes in their technical condition and monitoring of the beginning of defect origin.

Nowadays, noise is filtered in power station control systems in order to eliminate its effects on the results of analysis of technological parameters, resulting in the partial loss of measurement data obtained at the outputs of corresponding sensors in the latent period of an emergency state. For this reason, the use of traditional technologies for identifying changes in the technical condition of power engineering facilities at the early stage is ineffective, the results of diagnostics turn out to be belated, which, in turn, leads to disastrous accidents. We can assume that the disasters at the Chernobyl Nuclear Power Plant or Sayano–Shushenskaya Dam and others could have been prevented with the use of the technology of noise control of the beginning of the latent period of their transition into an emergency state. These examples indicate the significance of developing new technologies and systems of control and identification of the beginning of the latent period of changes in the technical condition of core modules of electric power stations [6–9].

Research shows that it is appropriate to use the technologies of correlation and spectral noise control of the beginning of microchanges in the technical condition of power engineering facilities to solve this problem. It is possible to design intelligent information systems based on these technologies for early detection of the latent

7.1 Specifics and Relevance of the Use of the Technology and System ...

period of defect initiation and accident prediction for these facilities. Therefore, their use can enhance the reliability and guarantee the safety of operation of hydroelectric and thermal power stations and nuclear power plants [6–14].

It is known [6–10] that the foundation of modern day thermal power station is powerful steam-turbine plants developed to a high level of excellence. They run on supercritical steam pressure and have advanced cycle diagram. For instance, combination plants using state-of-the-art high-temperature gas turbines have high thermal efficiency and energy conversion efficiency over 60%. The condition of these facilities is characterized by many technological parameters, e.g., for a steam-and-gas plant, namely, initial temperature of gas, intake air volume, pressure, capacity, turbine outlet temperature, high-pressure steam parameters, turbine, low-pressure steam parameters, energy conversion efficiency of the combination plant, specific reference fuel consumption, etc. The listed parameters contain sufficient information about the technical condition of the facility [1–4].

Control and management systems of thermal power stations, nowadays, monitor current technical condition by measuring a large number of technological parameters, including the aforementioned. Therefore, in addition to a wide range of functions mainly related to measurement, registration, and processing of various data, these systems are also responsible for the control of changes in the technical condition of these facilities. Of all functions of these systems, ensuring adequate and reliable identification of the technical condition and control of the beginning and development dynamics of accidents is of utmost importance. If we assume that a control object consists of n modules: $M_1, M_2, \ldots, M_n$, then the state of object Q_0 will be determined by the state of these modules, i.e.,

$$Q_0 = F(Q_{M_1}, Q_{M_1}, \ldots, Q_{M_p}).$$

Assume that the state of each of these modules is characterized by the input signals

$$g_{1M_1}(t), g_{2M_1}(t), \ldots, g_{iM_1}(t), \ldots, g_{mM_1}(t),$$
$$g_{1M_2}(t), g_{2M_2}(t), \ldots, g_{iM_2}(t), \ldots, g_{mM_2}(t),$$
$$\ldots$$
$$g_{1M_N}(t), g_{2M_N}(t), \ldots, g_{iM_N}(t), \ldots, g_{mM_N}(t),$$

and the output signals

$$\eta_{1M_1}(t), \eta_{2M_1}(t), \ldots, \eta_{iM_1}(t), \ldots, \eta_{mM_1}(t)$$
$$\eta_{1M_2}(t), \eta_{2M_2}(t), \ldots, \eta_{iM_2}(t), \ldots, \eta_{mM_2}(t),$$
$$\ldots$$
$$\eta_{1M_N}(t), \eta_{2M_N}(t), \ldots, \eta_{iM_N}(t), \ldots, \eta_{mM_N}(t).$$

To solve the problem of control of the technical condition of these modules, we first need to calculate the estimates of statistical characteristics of corresponding technological parameters. This allows for evaluating the state and quality of operation of modules over a particular period of time (e.g., a shift, a day, etc.). In [1–4] it is assumed that the implementations of $g(t)$ and $\eta(t)$ $\eta(t)$ of the technological parameters $X(t)$ and $Y(t)$ in power engineering are stationary ergodic with normal distribution law, and the selected time period T is fairly large, in consequence of which instead of the correlation functions $R_{XX}(\mu)$ and $R_{XY}(\mu)$ themselves of the useful signals $X(t)$ and $Y(t)$, the estimates $R_{gg}(\mu)$ and $R_{g\eta}(\mu)$ are usually used. They reflect the technical condition of the mentioned modules during their operation in the normal mode quite adequately. However, in the beginning of the latent period of facility's transition into an emergency state, the results of control based on these estimates often turns out to be belated. At the same time, in the beginning of defect origin, such estimates as noise variance and cross-correlation function between the noise and the useful signal change. Unfortunately, noises of technological parameters are filtered in monitoring and control systems of power engineering facilities, leading to the loss of important diagnostic information. This results in the missed opportunity of early monitoring of defect, on the one hand, and in a belated decision, on the other. Research shows that this specificity of control of the beginning of defect initiation at power engineering facilities is not accounted for in solving control and diagnostics problems [1–4]. This delays the results of control.

For this reason, there is currently a serious practical demand for technologies of noise control of the beginning of the latent period of transition of core modules of nuclear power plants, thermal, and hydroelectric power stations into an emergency state [1, 4].

An analysis of pre-accident situations at various power engineering facilities has demonstrated that when a facility's state is regarded as stable and the characteristics of the useful signal $X(t)$ are not changing, the estimates of characteristics of the noise $\varepsilon(t)$ that forms at the moment of the initial stage of appearance of various faults change abruptly at the beginning of the latent period of accidents [1, 3–6].

In view of the above, it is obvious that the development of intelligent systems for early detection of defect initiation based on noise technologies for core modules of power engineering facilities is of great strategic interest.

It is known that the technical condition of core modules of power engineering facilities is described by matrix equations of the type [1, 2]

$$\vec{r}_{g\eta}(\mu) \approx \vec{r}_{gg}(\mu)\vec{W}(\mu), \quad \mu = 0, \Delta t, 2\Delta t, \ldots, (N-1)\Delta t,$$

where $\vec{r}_{gg}(\mu)$ is the square symmetric matrix of the normalized autocorrelation functions with dimension $N \times N$ of the centered input signal $X(t)$; $\vec{r}_{g\eta}$ is the column vector of the normalized cross-correlation functions between the input $X(t)$ and the output $Y(t)$; $\overline{W}(\mu)$ is the column vector of the impulsive admittance functions. The normalized auto- and cross-correlation functions $r_{gg}(\mu), r_{g\eta}(\mu)$ of the noisy signals are calculated from the following formulas:

$$\left. \begin{array}{l} r_{gg}(\mu) \approx \frac{R_{gg}(\mu)}{D_g} \\ r_{g\eta}(\mu) \approx \frac{R_{g\eta}(\mu)}{\sqrt{D_g D_\eta}} \end{array} \right\}.$$

7.1 Specifics and Relevance of the Use of the Technology and System ...

However, in many cases, we fail to ensure adequacy of identification of the technical condition in the latent period of accidents by means of these matrices in practice [2, 5]. Accordingly, we need to form equivalent correlation matrices $\vec{r}^e_{gg}(\mu)$ and $\vec{r}^e_{g\eta}(\mu)$ that ensure that the following equalities hold:

$$\vec{r}^e_{gg}(\mu) \approx \vec{r}_{XX}(\mu),$$
$$\vec{r}^e_{g\eta}(\mu) \approx \vec{r}_{XY}(\mu).$$

In this case, taking into account the necessity to correct the errors of the normalization procedure, as indicated in Chap. 2, the formulas for calculating the normalized correlation functions have the following form:

$$r_{gg}(\mu \neq 0) \approx \frac{R_{gg}(\mu \neq 0)}{D_g - D_\varepsilon}, \qquad (7.1)$$

$$r_{g\eta}(\mu) \approx \frac{R_{g\eta}(\mu)}{\sqrt{(D_g - D_\varepsilon)(D_\eta - D_{\varphi\varphi})}}. \qquad (7.2)$$

Therefore, the normalized correlation matrix of the noisy signals $g(i\Delta t)$ for power engineering facilities can be written as

$$r^e_{gg}(\mu) \approx \begin{Vmatrix} 1 & \frac{R_{gg}(\Delta t) \approx R_{XX}(\Delta t)}{D_g - D_\varepsilon} & \cdots & \frac{R_{gg}[(N-1)\Delta t] \approx R_{XX}[(N-1)\Delta t]}{D_g - D_\varepsilon} \\ \frac{R_{gg}(\Delta t) \approx R_{XX}(\Delta t)}{D_g - D_\varepsilon} & 1 & \cdots & \frac{R_{gg}[(N-2)\Delta t] \approx R_{XX}[(N-2)\Delta t]}{D_g - D_\varepsilon} \\ \cdots & \cdots & \cdots & \cdots \\ \frac{R_{gg}[(N-1)\Delta t] \approx R_{XX}[(N-1)\Delta t]}{D_g - D_\varepsilon} & \frac{R_{gg}[(N-2)\Delta t] \approx R_{XX}[(N-2)\Delta t]}{D_g - D_\varepsilon} & \cdots & 1 \end{Vmatrix}. \qquad (7.3)$$

The matrix of the normalized cross-correlation functions can be formed in the similar manner:

$$\vec{r}^e_{g\eta}(\mu) \approx \begin{bmatrix} \frac{R_{g\eta}(0) \approx R_{XY}(0)}{\sqrt{(D_g - D_\varepsilon)(D_\eta - D_\varphi)}} & \frac{R_{g\eta}(\Delta t) \approx R_{XY}(\Delta t)}{\sqrt{(D_g - D_\varepsilon)(D_\eta - D_\varphi)}} & \cdots & \frac{R_{g\eta}[(N-1)\Delta t] \approx R_{XY}[(N-1)\Delta t]}{\sqrt{(D_g - D_\varepsilon)(D_\eta - D_\varphi)}} \end{bmatrix}^T. \qquad (7.4)$$

Obviously, through the use of formulas (7.1) and (7.2), we can, by eliminating said errors, form normalized correlation matrices (7.3) and (7.4), which are equivalent to the matrices of the useful signals [2, 3, 5].

It is essential to account for the correlation between $X(t)$ and $\varepsilon(t)$ when forming the correlation matrices because a correlation between $X(t)$ and $\varepsilon(t)$ often takes place during the dynamics of development of accidents at power engineering facilities even during several sampling intervals, i.e., at $\mu = \Delta t, \mu = 2\Delta t, \mu = 3\Delta t, \ldots$ [3, 4].

Therefore, to control development dynamics of accidents, we need to use technologies for calculating the estimates of the cross-correlation functions $R_{X\varepsilon}(0)$,

$R_{X\varepsilon}(\Delta t)$, $R_{X\varepsilon}(2\Delta t)$, $R_{X\varepsilon}(3\Delta t)$, …. In forming the correlation matrices, this allows ensuring their equivalence to the matrix of the useful signals by compensating for the errors of the elements $R_{gg}(0)$, $R_{gg}(\Delta t)$, $R_{gg}(2\Delta t)$, $R_{gg}(3\Delta t)$, … in the corresponding rows and columns of the correlation matrices (7.3). Thus, to ensure that the correlation matrices are equivalent to the matrices of the useful signals, we need to subtract the value of D_ε from the estimates of $R_{gg}(0) = D_g$, and the value of $R_{X\varepsilon}(\mu)$ from the estimates of $R_{gg}(\mu)$:

$$r_{gg}^e(\mu) \approx \begin{Vmatrix} 1 & \frac{R_{gg}(\Delta t)-R_{X\varepsilon}(\Delta t)}{D_g-D_\varepsilon} & \cdots & \frac{R_{gg}[(N-1)\Delta t]-R_{X\varepsilon}[(N-1)\Delta t]}{D_g-D_\varepsilon} \\ \frac{R_{gg}(\Delta t)-R_{X\varepsilon}(\Delta t)}{D_g-D_\varepsilon} & 1 & \cdots & \frac{R_{gg}[(N-2)\Delta t]-R_{X\varepsilon}[(N-2)\Delta t]}{D_g-D_\varepsilon} \\ \cdots & \cdots & \cdots & \cdots \\ \frac{R_{gg}[(N-1)\Delta t]-R_{X\varepsilon}[(N-1)\Delta t]}{D_g-D_\varepsilon} & \frac{R_{gg}[(N-2)\Delta t]-R_{X\varepsilon}[(N-2)\Delta t]}{D_g-D_\varepsilon} & \cdots & 1 \end{Vmatrix}.$$

Therefore, in forming the equivalent matrices, alongside with the calculation of the estimate of D_ε, it is also necessary to develop technologies for calculating the estimates of $R_{X\varepsilon}(1\Delta t)$, $R_{X\varepsilon}(2\Delta t)$, $R_{X\varepsilon}(3\Delta t)$, …, $R_{X\varepsilon}(m\Delta t)$ and correcting the corresponding elements of the correlation matrices. For instance, in the presence of a correlation between $X(t)$ and $\varepsilon(t)$ in the elements of $R_{gg}(\Delta t)$, $R_{gg}(2\Delta t)$, $R_{gg}(3\Delta t)$, they are corrected by subtracting from them the corresponding estimates of $R_{X\varepsilon}(\Delta t)$, $R_{X\varepsilon}(2\Delta t)$, $R_{X\varepsilon}(3\Delta t)$, and the value of D_ε in the columns and rows of the correlation matrices, in which they are located. Thus, we form the correlation matrices equivalent to the matrices of the useful signals, eliminating the difficulties of solving the problem of identification of the technical condition of the aforementioned modules. However, in most cases, it takes a lot of time. Therefore, in order to solve this problem in real time, in practice, it is advisable to carry out noise control of the beginning of initiation of various defects that subsequently cause the control objects to go into an emergency state.

7.2 Intelligent Robust System for Noise Control of the Latent Period of Transition of Power Engineering Facilities into an Emergency State

Accident-free operation of power engineering facilities requires determining the beginning of the latent period of accidents of their most vulnerable components during their functioning in normal mode [1, 2]. This, in turn, requires high sensitivity of informative attributes to all possible microchanges that precede serious accident situations at power engineering facilities. In view of this, the control system of each module should calculate not only the set of traditional estimates of input and output signals but also the set of estimates of noise characteristics of the corresponding technological parameters [3, 4]. As Fig. 7.1 shows, to control the beginning of the latent period of the transition of a facility into an emergency state with the use of the

7.2 Intelligent Robust System for Noise Control of the Latent …

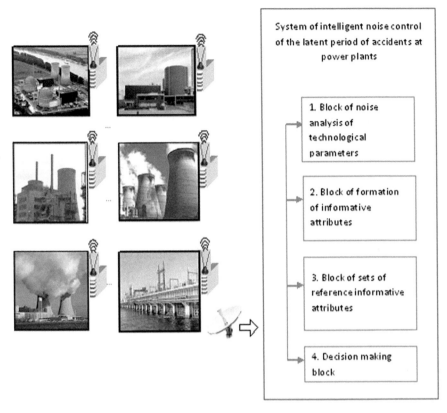

Fig. 7.1 Intelligent system of noise control of the latent period of accidents at power engineering facilities

intelligent technologies of noise control, the system operates at the first stage in the training mode. To this end, using the noise technologies presented in Chaps. 2–4, technological parameters are analyzed and the estimates of the noise characteristics of the analyzed signals $g_1(i\Delta t), g_2(i\Delta t), \ldots, g_j(i\Delta t)$ are calculated for each module in the normal technical condition. Further, from these estimates of the noise characteristics, reference combinations of informative attributes (CIA) are compiled, by means of which the set of reference informative attributes is formed

$$\begin{cases} D_{g_j\varepsilon}, R_{g_jX\varepsilon}(1\Delta t), R_{g_jX\varepsilon}(2\Delta t), R_{g_jX\varepsilon}(3\Delta t), \ldots, R_{g_jX\varepsilon}(m\Delta t) \\ R^*_{g_jX\varepsilon}(1\Delta t), R^*_{g_jX\varepsilon}(2\Delta t), R^*_{g_jX\varepsilon}(3\Delta t), \ldots, R^*_{g_jX\varepsilon}(m\Delta t) \\ a^R_{g_jn}, b^R_{g_jn}; a_{g_jn\varepsilon}, b_{a_jn\varepsilon}; a_{g_jn\varepsilon_2}, b_{y_jn\varepsilon_2} \\ a^*_{g_jn}, b^*_{g_jn}; a^*_{g_jn\varepsilon}, b^*_{g_jn\varepsilon}; a^*_{g_jn\varepsilon_2}, b^*_{g_jn\varepsilon_2} \\ k_{g_jq_0}, k_{g_jq_1}, k_{g_jq_2}, \ldots, k_{g_jq_m}; \bar{f}_{g_jq_0}, \bar{f}_{g_jq_1}, \bar{f}_{g_jq_2}, \ldots, \bar{f}_{g_jq_m} \end{cases}.$$

Algorithms and technologies for calculating and forming these informative attributes are discussed and described in detail in Chaps. 2–4.

During the operation of a facility in the normal mode, sets of reference informative attributes w_{g_1}, w_{g_2}, w_{g_j} form that correspond to the facility's normal technical condition

$$W_{g_1} = \begin{cases} D_{g_1\varepsilon}, R_{g_1 X\varepsilon}(1\Delta t), R_{g_1 X\varepsilon}(2\Delta t), R_{g_1 X\varepsilon}(3\Delta t), \ldots, R_{g_1 X\varepsilon}(m\Delta t) \\ R^*_{g_1 X\varepsilon}(1\Delta t), R^*_{g_1 X\varepsilon}(2\Delta t), R^*_{g_1 X\varepsilon}(3\Delta t), \ldots, R^*_{g_1 X\varepsilon}(m\Delta t) \\ a^R_{g_1 n}, b^R_{g_1 n}; a_{g_1 n\varepsilon}, b_{g_1 n\varepsilon}; a_{g_1 n\varepsilon_2}, b_{g_1 n\varepsilon_2} \\ a^*_{g_1 n}, b^*_{g_1 n}; a^*_{g_1 n\varepsilon}, b^*_{g_1 n\varepsilon}; a^*_{g_1 n\varepsilon_2}, b^*_{g_1 n\varepsilon_2} \\ k_{g_1 q_0}, k_{g_1 q_1}, k_{g_1 q_2}, \ldots, k_{g_1 q_m}; \bar{f}_{g_1 q_0}, \bar{f}_{g_1 q_1}, \bar{f}_{g_1 q_2}, \ldots, \bar{f}_{g_1 q_m} \end{cases},$$

$$W_{g_2} = \begin{cases} D_{g_2\varepsilon}, R_{g_2 X\varepsilon}(1\Delta t), R_{g_2 X\varepsilon}(2\Delta t), R_{g_2 X\varepsilon}(3\Delta t), \ldots, R_{g_2 X\varepsilon}(m\Delta t) \\ R^*_{g_2 X\varepsilon}(1\Delta t), R^*_{g_2 X\varepsilon}(2\Delta t), R^*_{g_2 X\varepsilon}(3\Delta t), \ldots, R^*_{g_2 X\varepsilon}(m\Delta t) \\ a^R_{g_2 n}, b^R_{g_2 n}; a_{g_2 n\varepsilon}, b_{g_2 n\varepsilon}; a_{g_2 n\varepsilon_2}, b_{g_2 n\varepsilon_2} \\ a^*_{g_2 n}, b^*_{g_2 n}; a^*_{g_2 n\varepsilon}, b^*_{g_2 n\varepsilon}; a^*_{g_2 n\varepsilon_2}, b^*_{g_2 n\varepsilon_2} \\ k_{g_2 q_0}, k_{g_2 q_1}, k_{g_2 q_2}, \ldots, k_{g_2 q_m}; \bar{f}_{g_2 q_0}, \bar{f}_{g_2 q_1}, \bar{f}_{g_2 q_2}, \ldots, \bar{f}_{g_2 q_m} \end{cases},$$

$$W_{g_j} = \begin{cases} D_{g_j\varepsilon}, R_{g_j X\varepsilon}(1\Delta t), R_{g_j X\varepsilon}(2\Delta t), R_{g_j X\varepsilon}(3\Delta t), \ldots, R_{g_j X\varepsilon}(m\Delta t) \\ R^*_{g_j X\varepsilon}(1\Delta t), R^*_{g_j X\varepsilon}(2\Delta t), R^*_{g_j X\varepsilon}(3\Delta t), \ldots, R^*_{g_j X\varepsilon}(m\Delta t) \\ a^R_{g_j n}, b^R_{g_j n}; a_{g_j n\varepsilon}, b_{g_j n\varepsilon}; a_{g_j n\varepsilon_2}, b_{g_j n\varepsilon_2} \\ a^*_{g_j n}, b^*_{g_j n}; a^*_{g_j n\varepsilon}, b^*_{g_j n\varepsilon}; a^*_{g_j n\varepsilon_2}, b^*_{g_j n\varepsilon_2} \\ k_{g_j q_0}, k_{g_j q_1}, k_{g_j q_2}, \ldots, k_{g_j q_m}; \bar{f}_{g_j q_0}, \bar{f}_{g_j q_1}, \bar{f}_{g_j q_2}, \ldots, \bar{f}_{g_j q_m} \end{cases}.$$

These sets are entered into the block of sets of reference informative attributes (BSRIA), which is individual for each module. The duration of the training process depends on the specific characteristics of facilities [4, 5]. In the future, if any combination of estimates matches one already present in the BSRIA, the decision-making block (DMB) registers the fact of match. If a current combination differs from all those present in the BSRIA, a new reference is included in the corresponding set. At some point, newly obtained current estimates repeatedly turn out to be equal or close in value to the combinations in the corresponding set in the BSRIA. The first stage of the training process is regarded as complete only after such results have been obtained on multiple occasions [4, 5]. Thus, BSRIA forms at the first stage from the estimates of the noise characteristics, allowing one to perform control of the beginning of the latent period of accidents.

At the second stage of the process of control of facility's technical condition, the calculation of informative attributes continues. If a current combination of estimates matches the estimates from the corresponding set already present in the BSRIA, this is regarded as a confirmation of the facility's normal operation. If a current combination in some time period differs from all those present in the BSRIA, DMB forms a message about the start of possible microchanges in the technical condition of the facility and alerts the operating personnel, also providing the list of all combinations of estimates from the BSRIA that correspond to that condition, as well as the esti-

7.2 Intelligent Robust System for Noise Control of the Latent …

mates from the current combination. If no malfunction is detected after appropriate thorough examination of the facility's technical condition, and it is established that the facility operates normally, the current combination of informative attributes is also included in the corresponding set of reference combinations and saved in the BSRIA. Otherwise, the beginning of the transition of the facility into an emergency state is recorded. In that case, an appropriate message forms in the DMB, which is registered as a report indicating the time and results of monitoring [1–5].

In conclusion, we should note that during the training in these subsystems in the process of noise control of a facility under its normal operating mode, it is possible to determine the combinations of the maximum threshold values of the estimates of various noise characteristics for all analyzed signals. After a certain period of time, from these combinations of various maximum estimates, a reference set corresponding to the facility's normal technical condition is formed and saved. It has the following form:

$$W_j^{\max} = \begin{cases} D_{g_j\varepsilon}^{\max}, R_{g_jX\varepsilon}^{\max}(1\Delta t), R_{g_jX\varepsilon}^{\max}(2\Delta t), R_{g_jX\varepsilon}^{\max}(3\Delta t), \ldots, R_{g_jX\varepsilon}^{\max}(m\Delta t) \\ R_{g_jX\varepsilon}^{*\max}(1\Delta t), R_{g_jX\varepsilon}^{*\max}(2\Delta t), R_{g_jX\varepsilon}^{*\max}(3\Delta t), \ldots, R_{g_jX\varepsilon}^{*\max}(m\Delta t) \\ a_{g_jn}^{\max}, b_{g_jn}^{\max}; a_{g_jn\varepsilon}^{\max}, b_{g_jn\varepsilon}^{\max}; a_{g_jn\varepsilon_2}^{\max}, b_{g_jn\varepsilon_2}^{\max} \\ a_{g_jn}^{*\max}, b_{g_jn}^{*\max}; a_{g_jn\varepsilon}^{*\max}, b_{g_jn\varepsilon}^{*\max}; a_{g_jn\varepsilon_2}^{*\max}, b_{g_jn\varepsilon_2}^{*\max} \\ k_{g_jq_0}^{\max}, k_{g_jq_1}^{\max}, k_{g_jq_2}^{\max}, \ldots, k_{g_jq_m}^{\max}; \bar{f}_{g_jq_0}^{\max}, \bar{f}_{g_jq_1}^{\max}, \bar{f}_{g_jq_2}^{\max}, \ldots, \bar{f}_{g_jq_m}^{\max} \end{cases}.$$

When the subsystem is operating in the noise control mode during the operation of the equipment, the obtained current combination of estimates of various noise characteristics of all analyzed signals is compared with the corresponding maximum reference threshold estimates. The moment at which the current estimates become greater or equal to the corresponding reference threshold estimates is taken as the beginning of the latent period of an emergency state.

Thus, the use of the intelligent noise control system shown in Fig. 7.1 makes it possible to continuously control the beginning of the latent period and the dynamics of the development of accidents of the core modules of power stations. This allows raising the level of safety and reliability of operation of power engineering facilities.

7.3 Subsystems for Noise Control of the Beginning of the Latent Period of Nuclear Power Plant Equipment's Transition into an Emergency State

Modern systems of online diagnostics are widely used at nuclear power plants (NPP) regardless of the type of reactor. For instance, Fig. 7.2 shows a system for online diagnostics of NPP equipment with a WWER-1000 reactor. The use of the system does not depend on the type of NPP reactor [6–14].

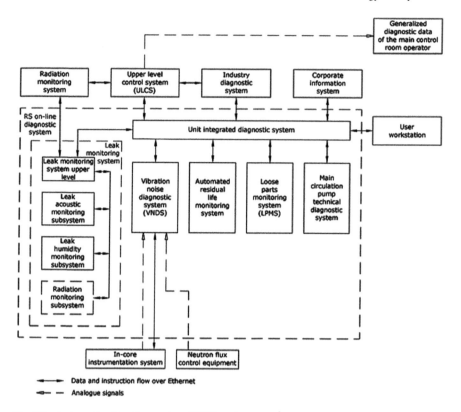

Fig. 7.2 System of online diagnostics of NPP equipment

As can be seen from the diagram, the complex diagnostic system consists of the following:

(1) vibration noise diagnostics system (VNDS);
(2) automated residual life monitoring systems (ARLMS);
(3) loose parts monitoring system (LPMS);
(4) main circulation pump technical diagnostic system (MCP TDS);
(5) radiation monitoring system.

The radiation monitoring system consists of the following:

(1) leak acoustic monitoring subsystem (AMS);
(2) leak humidity monitoring subsystem (HMS);
(3) radiation monitoring control (RMC).

The combination of these subsystems provides control and diagnostics of NPP equipment [9, 14].

It should be noted that this system of online diagnostics of NPP equipment with WWER-1000 reactor has been in operation at Units 1, 2 of Tianwan NPP (China),

7.3 Subsystems for Noise Control of the Beginning of the Latent ...

installed at Units 1, 2 of Kudankulam NPP (India), Unit 1 of Bushehr NPP (Iran), Units 3, 4 of Kalinin NPP, Units 2, 3 of Rostov NPP, Unit 6 of Novovoronezh NPP, Units 3, 4 of Kola NPP, Unit 1 of Leningrad NPP-2 (Russia).

The systems operate on the basis of the most advanced modern noisy signal analysis technologies. However, they cannot analyze the noises that emerge at the beginning of the latent period of the initiation of defects preceding the crashes. For this reason, at the beginning of the latent period of initiation of possible defects, these systems respond with a delay. In addition, these systems cannot control the dynamics of the development of an accident in its latent period [6–14].

In view of the above, in order to improve the effectiveness of control of the beginning and development dynamics in the latent period of faults, it is advisable to equip these systems with subsystems that allow performing noise control of initiation of defects that subsequently lead to an emergency state of NPP equipment. These subsystems operate on the basis of noise technologies for analyzing measurement information by means of appropriate software tools. Thus, in the existing system of complex diagnostics, the reliability and validity of the control of the beginning of changes in the latent period of the emergency state of NPP equipment are increased. At the same time, the noise of the measuring information received at the output of the existing sensors is used as the carrier of diagnostic information, and for this purpose, the system of online diagnostics of NPP equipment is equipped with the following new noise technologies-based subsystems:

- Vibration noise diagnostics system (VNDS) with the subsystem of noise monitoring of the beginning of microvibration.
- Automated residual life monitoring systems (ARLMS) with the subsystem of noise control of residual life dynamic.
- Loose parts monitoring system (LPMS) with the subsystem of noise monitoring of the beginning of failure in metal structures.
- Main circulation pump technical diagnostic system (MCP TDS) with the subsystem of noise monitoring of the beginning of the latent period of MCP transition into an emergency state.

The subsystems shown in Fig. 7.3 increase the reliability and validity of control of the beginning of the latent period of faults in the relevant equipment. They are an integral part of the system of online monitoring of the technical condition of NPP equipment.

The relevance of the proposed variant of connecting additional noise control subsystems to the system of operational diagnostics is due to the possibility of increasing the safety level of NPP operation by using noise technologies to control the beginning of the latent period of equipment's transition into an emergency state. The use of technologies for calculating the noise estimates of the characteristics of noisy signals will allow early detection of changes in the technical condition of the corresponding units of NPP equipment. The reason is that at the initial stage of the latent period of defect initiation, noise arises in the signals received at the outputs of the sensors of NPP equipment, which correlates with the useful signal. In the proposed subsystems, the characteristics of this noise are used as carriers of diagnostic information.

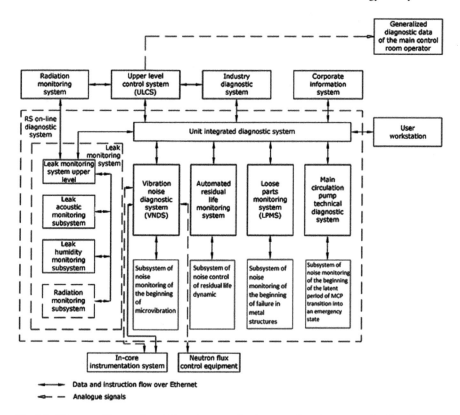

Fig. 7.3 Diagram of integration of the noise control subsystems into the system of online diagnostics of NPP equipment

As was mentioned above, algorithms and software of noise analysis of the measurement information received at the outputs of sensors installed on the NPP equipment are used for this purpose.

Studies have shown [1–4] that the use of the following noise control technologies is advisable:

1. technology of correlation noise control of the technical condition of control objects;
2. technology of spectral noise analysis;
3. position-binary technology of analysis of cyclic noisy signals;
4. technology of correlation noise analysis;
5. technology of spectral relay noise analysis;
6. technology of control of accident development dynamics.

All the above technologies have passed evaluation tests in systems of noise control of the beginning of changes in the technical condition of various facilities, and their efficiency has been proved operationally [3–5].

7.3 Subsystems for Noise Control of the Beginning of the Latent ...

Thus, during the operation of NPP equipment, the appearance of various defects in the form of wear, microcracks, carbon formation, fatigue deformation, etc., lead to the latent period of changes in their technical condition. This manifests itself as the noises $\varepsilon_{21}(i\Delta t), \varepsilon_{22}(i\Delta t), \varepsilon_{23}(i\Delta t), \ldots, \varepsilon_{2m}(i\Delta t)$ correlated with the useful signals $X(i\Delta t), X_2(i\Delta t), X_3(i\Delta t), \ldots, X_m(i\Delta t)$ of the noisy signals $g_1(i\Delta t), g_2(i\Delta t), g_3(i\Delta t), \ldots, g_m(i\Delta t)$. And this, in turn, leads to a change in the estimates of the above noise characteristics. Because of this, the system shown in Fig. 7.3 can control the beginning of the latent period of transition of NPP equipment into an emergency state by means of the set of specified informative noise attributes [6–14].

For instance, as experimental studies have shown, when monitoring the beginning of accidents at compressor stations, oil production facilities, drilling rigs, offshore fixed platforms, etc., the deviation of the current combinations of these informative attributes from the corresponding estimates of the elements of the reference sets made it possible to reliably register the beginning of the transition of these facilities from the normal state to an emergency state [1–4].

The subsystems proposed in Fig. 7.3 at the first stage operate in the mode of control of the beginning of the latent period of facility's transition to an emergency state in the training mode. In this case, first of all, reference combinations of informative attributes (RIA) are compiled from the noise estimates of the noise characteristics of the corresponding signals.

Further, sets of reference informative attributes (SRIA) corresponding to facility's normal technical condition are formed from these combinations of reference informative attributes.

$$W_{g_j} = \begin{cases} D_{g_j\varepsilon}, R_{g_jX\varepsilon}(1\Delta t), R_{g_jX\varepsilon}(2\Delta t), R_{g_jX\varepsilon}(3\Delta t), \ldots, R_{g_jX\varepsilon}(m\Delta t) \\ R^*_{g_jX\varepsilon}(1\Delta t), R^*_{g_jX\varepsilon}(2\Delta t), R^*_{g_jX\varepsilon}(3\Delta t), \ldots, R^*_{g_jX\varepsilon}(m\Delta t) \\ a^R_{g_jn}, b^R_{g_jn}; a_{g_jn\varepsilon}, b_{g_jn\varepsilon}; a_{g_jn\varepsilon_2}, b_{g_jn\varepsilon_2} \\ a^*_{g_jn}, b^*_{g_jn}; a^*_{g_jn\varepsilon}, b^*_{g_jn\varepsilon}; a^*_{g_jn\varepsilon_2}, b^*_{g_jn\varepsilon_2} \\ k_{g_jq_0}, k_{g_jq_1}, k_{g_jq_2}, \ldots, k_{g_jq_m}; \bar{f}_{g_jq_0}, \bar{f}_{g_jq_1}, \bar{f}_{g_jq_2}, \ldots, \bar{f}_{g_jq_m} \end{cases}.$$

The duration of the training process depends on the specific features of the functioning of the facility. At the same time, if any combination of estimates during the training process matches one of those already available in the SRIA, the fact of the match is registered in the decision-making block. If the current combination is different from all those already available in the SRIA, a new reference is entered in the corresponding set. At some point, newly obtained current estimates repeatedly turn out to be equal or close in value to certain combinations from the corresponding set in the SRIA. The first stage of the training process is regarded as complete only after such results have been obtained on multiple occasions. Thus, at the first stage, the knowledge base (KB) forms, which allows performing the control of the beginning of defect initiation [1–4].

At the second stage, the control of the beginning of the latent period of accidents begins. The calculation of the informative attributes and the formation of RIA from them is carried out in the same manner as described earlier. However, in this case, if a current combination of estimates matches the estimates already present in the SRIA from the set corresponding to the normal condition, this is taken as the confirmation of the normal technical condition. If a current combination in some time period differs from all sets of the normal condition, then information about possible microchanges in the technical condition in the corresponding section of equipment is formed. At the same time, this information is sent via a radio modem to the monitor of the operating personnel.

The personnel is also provided with the list of all combinations of estimates from the SRIA that correspond to that condition, as well as the values of the estimates from the current combination. If no malfunction is detected after the operating personnel carries out appropriate measures of diagnostics and control of facility's technical condition, and it is established that the facility operates in the normal mode, the current RIA is also entered in the corresponding set of reference combinations and saved in the SRIA. Otherwise, the beginning of the transition of the equipment into an emergency state is recorded, the appropriate information is formed, which is registered as a report indicating the time and results of monitoring and also displayed at the system monitor.

In conclusion, it should be noted that the process of noise control in these subsystems can also be performed by determining the maximum threshold values of the estimates of the various noise characteristics for all analyzed signals. To this end, during the training period, from the combination of the various maximum noise estimates, a reference set of the following form is created and saved

$$W_j^{\max} = \begin{cases} D_{g_j\varepsilon}^{\max}, R_{g_jX\varepsilon}^{\max}(1\Delta t), R_{g_jX\varepsilon}^{\max}(2\Delta t), R_{g_jX\varepsilon}^{\max}(3\Delta t), \ldots, R_{g_jX\varepsilon}^{\max}(m\Delta t) \\ R_{g_jX\varepsilon}^{*\max}(1\Delta t), R_{g_jX\varepsilon}^{*\max}(2\Delta t), R_{g_jX\varepsilon}^{*\max}(3\Delta t), \ldots, R_{g_jX\varepsilon}^{*\max}(m\Delta t) \\ a_{g_jn}^{\max}, b_{g_jn}^{\max}; a_{g_jn\varepsilon}^{\max}, b_{g_jn\varepsilon}^{\max}; a_{g_jn\varepsilon_2}^{\max}, b_{g_jn\varepsilon_2}^{\max} \\ a_{g_jn}^{*\max}, b_{g_jn}^{*\max}; a_{g_jn\varepsilon}^{*\max}, b_{g_jn\varepsilon}^{*\max}; a_{g_jn\varepsilon_2}^{*\max}, b_{g_jn\varepsilon_2}^{*\max} \\ k_{g_jq_0}^{\max}, k_{g_jq_1}^{\max}, k_{g_jq_2}^{\max}, \ldots, k_{g_jq_m}^{\max}; \bar{f}_{g_jq_0}^{\max}, \bar{f}_{g_jq_1}^{\max}, \bar{f}_{g_jq_2}^{\max}, \ldots, \bar{f}_{g_jq_m}^{\max} \end{cases}.$$

Due to this, during the operation of the equipment, the obtained current combination of estimates of the various noise characteristics of all analyzed signals is compared with the corresponding maximum reference threshold estimates. The moment at which current estimates become greater or equal to the corresponding reference threshold estimates is taken as the beginning of the latent period of an emergency state.

7.4 Possibility of Preflight Noise Control of the Technical Condition of Aviation Equipment

Most failures of aviation equipment are preceded by the initiation of some specific defects that are caused various operational factors and arise in the form of wear, cracks, fatigue deformation, overload, etc. The effectiveness of technical operation of aviation equipment largely depends on the validity of the control of the beginning and development dynamics of the latent period of an emergency state of aircrafts and their engines. It is known that new generations of control and management systems using all kinds of advanced methods and technologies of control, diagnostics, and identification have been created in recent years [1–4].

For instance, in the diagnostics of the technical condition of gas turbine engines, technologies of spectral analysis of vibration signals are widely used. Characteristic for this method of control is a large number of vibration sensors (from 40 to 100 pcs) placed in all vulnerable elements of an aircraft. This method suggests acquisition of measurement information received from vibration sensors located on a large number of various elements of the object. Control and identification are carried out through spectral analysis of signals and calculation of current informative attributes. For this purpose, they are compared with the reference estimates of the spectral characteristics of vibration formed during the training. The vibration level depends on many factors, such as operating principle of the element, method of application of the vibration sensors and the level of dynamic forces and moments acting in them, their distance from the vibration source, location and method of mounting of the vibration sensors, characteristics of the vibration sensor itself, etc., as well as the time and location of the aircraft and the environment. The system allows determining successively changing vibration in certain frequency intervals for all elements of the object and making an appropriate decision. However, the shortcoming of this technology is that it cannot ensure the reliability of control of the beginning of defect initiation. This is because all signals that come from vibration sensors to form the reference base and to identify the current signal are subjected to thorough decontamination the sum noise $\varepsilon(i\Delta t)$. This leads to a loss of diagnostic information. Therefore, only the obvious faults are diagnosed in this case. For this reason, the control system does not respond to microdamage, which in the event of an unsuccessful combination of factors during flight can lead to a plane crash.

This drawback is also characteristic of other methods and technologies of analysis of measurement information widely used in control systems. Therefore, despite the use of many advanced technologies in recent years and the increased reliability of the element base of these systems, the number of accidents still remains unreasonably high.

Because of these specifics of aviation equipment, it is obvious that it is necessary to control defect initiation at the stage of micro- and minichanges in the state of the most vulnerable components. This, in turn, requires a high sensitivity of informative attributes to all kinds of microchanges, which in airliners usually precede serious emergency situations. From this point of view, in addition to traditional estimates, it is also appropriate to ensure the calculation of the set of noise estimates of the characteristics of noisy signals in order to improve the reliability and validity of the control results. For this purpose, existing control systems can be equipped with subsystems of correlation and spectral noise control of the beginning and development dynamics of the latent period of accidents. Due to this, by duplicating the traditional and several noise technologies, it is possible to ensure sufficient reliability of the control results.

An analysis of the possibilities of using noise control on airplanes has shown that, based on this technology, it is expedient to create a system of preflight noise control of the initiation of faults in aircraft's technical condition. For the purpose of the preflight control, vibrating sensors should be located in the vulnerable parts of its fuselage. During the flight preparation, vibration signals can be analyzed in conditions determined by the safety requirements. In this case, the beginning of the latent period of an emergency state manifests itself in the noises $\varepsilon(i\Delta t) = \varepsilon_1(i\Delta t) + \varepsilon_2(i\Delta t)$ of the vibration signals:

$$g_1(i\Delta t), g_2(i\Delta t), g_3(i\Delta t), \ldots, g_i(i\Delta t), \ldots, g_m(i\Delta t).$$

By calculating the estimates of their noise characteristics according to the technologies described in Chaps. 2–4, we control the beginning of the initiation of faults [1–4]. For this purpose, the sets W_{g_j} of the informative attributes consisting of the estimates of the noise variance, cross-correlation functions between the useful signal and the noise, etc., from the obtained estimates of every signal $g_j(i\Delta t)$ are formed in the following form:

$$W_{g_j} = \begin{cases} D_{g_j\varepsilon}, R_{g_jX\varepsilon}(1\Delta t), R_{g_jX\varepsilon}(2\Delta t), R_{g_jX\varepsilon}(3\Delta t), \ldots, R_{g_jX\varepsilon}(m\Delta t) \\ R^*_{g_jX\varepsilon}(1\Delta t), R^*_{g_jX\varepsilon}(2\Delta t), R^*_{g_jX\varepsilon}(3\Delta t), \ldots, R^*_{g_jX\varepsilon}(m\Delta t) \\ a^R_{g_jn}, b^R_{g_jn}; a_{g_jn\varepsilon}, b_{g_jn\varepsilon}; a_{g_jn\varepsilon_2}, b_{g_jn\varepsilon_2} \\ a^*_{g_jn}, b^*_{g_jn}; a^*_{g_jn\varepsilon}, b^*_{g_jn\varepsilon}; a^*_{g_jn\varepsilon_2}, b^*_{g_jn\varepsilon_2} \\ k_{g_jq_0}, k_{g_jq_1}, k_{g_jq_2}, \ldots, k_{g_jq_m}; \bar{f}_{g_jq_0}, \bar{f}_{g_jq_1}, \bar{f}_{g_jq_2}, \ldots, \bar{f}_{g_jq_m} \end{cases}$$

During the operation of the system at the training stage, the reference sets of estimates of informative attributes in the form of a group of the indicated sets W_{g_1}, W_{g_2}, and W_{g_j} are obtained for all necessary preflight situations of aircraft operation.

In the period T_0, when the object is in the normal state, all current informative attributes are within the reference range. At the moment of defect origin, the system perceives a deviation of any element of the sets from the reference range W_{g_1}, W_{g_2}, and W_{g_j} as the beginning of the transition of the control object from the period of the normal technical condition into the period of a latent emergency state. The number of the set, the number of the column and the row of the location of the element of the informative attribute that falls outside the reference range can be used to identify the location and nature of the fault. A deviation of any element from the range of the corresponding reference elements of the specified set indicates an emergency state of the corresponding element of the object. Combining (duplicating) the control by means of many different informative attributes makes it possible to significantly increase the reliability of the control results.

The effectiveness of noise control by the proposed method is it is quite easy to calculate the noise estimates of each of the specified informative attributes, and the results are reliable. In addition, the use of combinations of the indicated noise technologies of analysis of measurement information increases the reliability of preflight control of aviation equipment, and the control method itself can also be used to

7.4 Possibility of Preflight Noise Control of the Technical ...

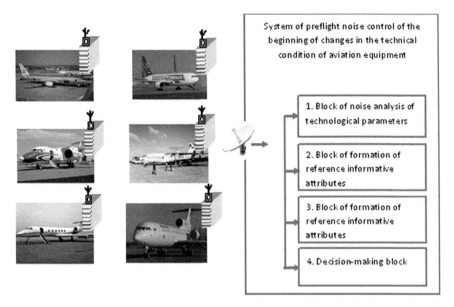

Fig. 7.4 Intelligent system of preflight control of the beginning of changes in the technical condition of aviation equipment

monitor the technical condition of any aircraft subjected to various dynamic loads [2, 3]. Considering the catastrophic consequences of numerous aircraft accidents, the expediency of their use in practice is obvious. One of the possible variants of preflight noise control systems is shown in Fig. 7.4. In this system for the preflight control of the beginning of the latent period of aircraft's transition into an emergency state, intelligent technology of noise control is used, which at the first stage works in the training mode. For this, using the vibration signals $g_j(i\Delta t)$ in the initial technical condition of the i-th element of the fuselage, first of all, reference combinations of informative attributes (RIA) are compiled from the noise estimates of the characteristics of technological parameters [3–5].

Further, sets of reference informative attributes (SRIA) W_i corresponding to aircraft's normal technical condition are formed from these combinations of reference informative attributes (RIA). For this, from the various noise characteristics during the training process, reference sets are formed and saved in the following form:

$$W_{g_j}^{\max} = \begin{cases} D_{g_j\varepsilon}, R_{g_jX\varepsilon}(1\Delta t), R_{g_jX\varepsilon}(2\Delta t), R_{g_jX\varepsilon}(3\Delta t), \ldots, R_{g_jX\varepsilon}(m\Delta t) \\ R^*_{g_jX\varepsilon}(1\Delta t), R^*_{g_jX\varepsilon}(2\Delta t), R^*_{g_jX\varepsilon}(3\Delta t), \ldots, R^*_{g_jX\varepsilon}(m\Delta t) \\ a^R_{g_jn}, b^R_{g_jn}; a_{g_jn\varepsilon}, b_{g_jn\varepsilon}; a_{g_jn\varepsilon_2}, b_{g_jn\varepsilon_2} \\ a^*_{g_jn}, b^*_{g_jn}; a^*_{g_jn\varepsilon}, b^*_{g_jn\varepsilon}; a^*_{g_jn\varepsilon_2}, b^*_{g_jn\varepsilon_2} \\ k_{g_jq_0}, k_{g_jq_1}, k_{g_jq_2}, \ldots, k_{g_jq_m}; \bar{f}_{g_jq_0}, \bar{f}_{g_jq_1}, \bar{f}_{g_jq_2}, \ldots, \bar{f}_{g_jq_m} \end{cases}.$$

These sets are entered in the SRIA, which is individual for each aircraft. The duration of the training process depends on the specifics the aviation equipment. In the future, if any combination of current estimates matches one of those already available in the SRIA, the fact of the match is registered in the decision-making block (DMB). If the current combination is different from all those already available in the SRIA, a new reference is entered in the corresponding set. At some point, newly obtained current maximum estimates of the noise characteristics repeatedly turn out to be equal or close in value to combinations from the corresponding set in the SRIA. The first stage of the training process is regarded as complete only after such results have been obtained on multiple occasions. Thus, at the first stage, the SRIA forms, which allows performing the monitoring of the beginning of changes in the technical condition of aviation equipment.

At the second stage of the monitoring process, the calculation of the informative attributes and the formation of RIA from them is continued. In this case, if a current combination of estimates does not exceed the estimates already present in the SRIA from the corresponding set, this is taken as the confirmation of the normal operation of the aircraft. If a current combination in some time period will have a smaller value than all the reference threshold estimates already present in the SRIA, then information about possible microchanges in the technical condition in the aviation equipment is formed in DMB and sent to the operating personnel. The personnel is also provided with the list of all combinations of estimates from the SRIA that correspond to that condition, as well as the values of the estimates of the current combination. If no malfunction is detected after appropriate measures of thorough control of aircraft's technical condition, and it is established that it functions normally, the current RIA is also entered in the corresponding set of reference combinations and saved in the SRIA. Otherwise, the beginning of the transition of the aviation equipment into an emergency state is recorded. In this case, the DMB forms the relevant information, which is registered as a report indicating the time and results of monitoring.

In conclusion, it should be noted that it is possible to accelerate the training process and simplify the formation of sets of reference informative attributes. To this end, during the training process, the maximum value of the corresponding informative attributes is determined for the normal technical condition of the aircraft in the following form:

$$W_j^{\max} = \begin{cases} D_{g_j\varepsilon}^{\max}, R_{g_jX_\varepsilon}^{\max}(1\Delta t), R_{g_jX_\varepsilon}^{\max}(2\Delta t), R_{g_jX_\varepsilon}^{\max}(3\Delta t), \ldots, R_{g_jX_\varepsilon}^{\max}(m\Delta t) \\ R_{g_jX_\varepsilon}^{*\max}(1\Delta t), R_{g_jX_\varepsilon}^{*\max}(2\Delta t), R_{g_jX_\varepsilon}^{*\max}(3\Delta t), \ldots, R_{g_jX_\varepsilon}^{*\max}(m\Delta t) \\ a_{g_jn}^{\max}, b_{g_jn}^{\max}; a_{g_jn\varepsilon}^{\max}, b_{g_jn\varepsilon}^{\max}; a_{g_jn\varepsilon_2}^{\max}, b_{g_jn\varepsilon_2}^{\max} \\ a_{g_jn}^{*\max}, b_{g_jn}^{*\max}; a_{g_jn\varepsilon}^{*\max}, b_{g_jn\varepsilon}^{*\max}; a_{g_jn\varepsilon_2}^{*\max}, b_{g_jn\varepsilon_2}^{*\max} \\ k_{g_jq_0}^{\max}, k_{g_jq_1}^{\max}, k_{g_jq_2}^{\max}, \ldots, k_{g_jq_m}^{\max}; \bar{f}_{g_jq_0}^{\max}, \bar{f}_{g_jq_1}^{\max}, \bar{f}_{g_jq_2}^{\max}, \ldots, \bar{f}_{g_jq_m}^{\max} \end{cases}.$$

In this variant, after the formation of the set of maximum reference informative attributes, if the current estimate of any informative attributes during the preflight control turns out to be greater than the corresponding maximum reference estimate, this is taken as the beginning of a latent emergency state of the aircraft.

7.5 Possibilities of Using Noise Control Technology in Railway Safety Systems in Seismically Active Regions

It is a fact that rail transport is currently the main mode of transport, both internationally and domestically. In many countries and regions, the primary transportation load is borne by rail transport [15, 16]. For this, massive railway communications have been created with modern bridges, tunnels, and stations. Management of this gigantic and complex economy is carried out by a variety of modern sophisticated control and diagnostic systems that ensure a high level of efficiency and safety of the rail transport [1, 2]. However, despite this, accidents of both freight and passenger trains are frequent. Therefore, as our studies have shown, by solving certain problems, such as ensuring the control of the latent period of technical condition of the railway tracks, railway bridges, tunnels, crossings, individual cars, the entire rolling stock, etc., it is possible to improve the safety of rail transport [3, 4]. This is especially important for rail transport in countries located in seismically active regions. This is due to the fact that weak (1–3 points) earthquakes occur in these regions quite often, affecting the technical condition of the railway tracks, bridges, tunnels, and communications. Not resulting in serious destruction, each of such earthquake is nevertheless a potential factor contributing to the beginning of the latent period of changes in the technical condition of various facilities. In this regard, the use of the technology and the noise control system in railway safety systems in seismically active regions and countries is of apparent practical interest [2–4].

A prerequisite for achieving high competitiveness in the transport market is the meeting the growing demand for speed, safety, and comfort. The significant advantages of rail transport in comparison with other modes of transport open up great prospects for increasing the volume of traffic and at the same time require improving this mode of transport [15, 16]. At present, there are great developments in this field, however, the specific conditions of seismically active regions create problems. The most important and vulnerable elements of rail transport are the moving part of cars and the railway tracks, and the research gap in the area of their interaction causes problems in achieving safety of the movement speed of the rolling stock [15, 16]. When a railway passes through seismically active regions, there are additional requirements to traffic safety. This particularly applies to the problems of control of the beginning of an emergency state of wheels and rails, their speed, derailing conditions.

Our analysis of literature [1, 2] on control technologies and systems taking into account the peculiarities of the railways in seismically active regions has shown that the use of noise technologies due to their specifics can indeed enhance the safety of this mode of transport. For this purpose, it is expedient to create subsystems of noise warning of seismic hazard, subsystems of noise control of the beginning, and development dynamics of changes in the technical condition of the rolling stock and tracks of the railway.

In view of the above, the issues of the technical safety of the rolling stock and the railway tracks with allowance for the specifics of the seismically active regions along the route, comprise an extremely topical problem. Studies have shown that the use of noise control technology in rail transport requires developing the following:

- an intelligent seismic hazard warning system throughout the main line;
- a subsystem of control of the beginning of the latent period of the technical condition of individual cars and the entire rolling stock;
- a subsystem of noise control of malfunctions on railway tracks, bridges, tunnels and communications throughout the entire route.

These subsystems will allow the driver and the dispatch service to obtain in advance additional information that makes it possible to improve traffic safety in general by taking appropriate measures.

It is obvious from the above that the use of noise control technologies in railway safety systems is best to be implemented in the following directions [1–4].

1. **Developing an intelligent seismic-acoustic seismic hazard warning network.** This involves creating a network of seismic-acoustic stations along the route, as well as developing the software and equipment for the transmission, acquisition, processing, and analysis of seismic-acoustic information. This subsystem, by means of a network of seismic-acoustic stations (the structural principle of which is given in Chap. 6), on the basis of noise analysis of seismic-acoustic signals makes it possible to calculate the probability of occurrence of dangerous earthquakes in the territory of the railway route in real time. The obtained information can be used in the relevant decision-making.

2. **Developing noise controllers with vibration sensors, which will allow to control the beginning and dynamics of changes in the technical condition of railway cars.** For this purpose, vibrating sensors should be installed in the most informative elements of controlled cars and other parts of the rolling stock. Due to this, the information obtained from each noise controller, which will reflect the technical condition of each car and the rolling stock as a whole, can be transmitted to the interface of the rolling stock control system at a distance of at least 500–600 m.

3. **Subsystem of noise control of malfunctions on railway tracks, bridges, tunnels and nodal communications of the whole route.** For this purpose, vibration sensors are installed on the most informative elements of the frame of the bridge or tunnel, which, by means of noise controllers, form the information reflecting the technical condition of railway cars at the time when they move across a particular section of the road.

The combination of these subsystems can contribute to improving the effectiveness of rail transport safety systems. All the proposed algorithms, technologies and

7.5 Possibilities of Using Noise Control Technology in Railway …   141

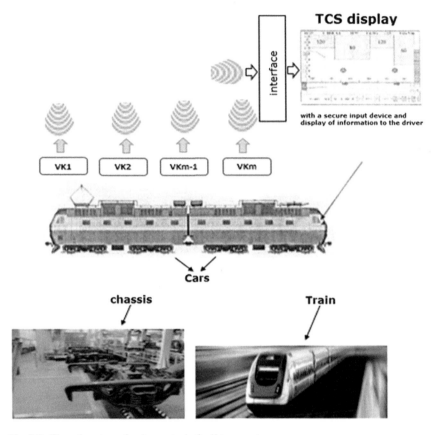

Fig. 7.5 The subsystem of noise control of rail transport

systems have been tested on other technical facilities, which gives grounds to believe that the proposed subsystems can be implemented in real conditions.

Figure 7.5 demonstrates the structural principle of one of the possible variants of the subsystem of noise control of rail transport. In this variant, Accutech AM20 vibration sensors are installed under the chassis wheels of each car. Vibration signals from these sensors are sent to the input of the controller of this car via radio channel. Any change in the state (deviation from the norm) of rails or tracks during the movement of the train manifests itself in the vibration sensor signals, which are analyzed on the noise analysis controllers. Due to this, the beginning of possible malfunctions at this moment affects the estimate of the noise characteristics of the vibration signal [1–4].

On the noise controllers, these signals are analyzed, for instance, on the basis of the relay correlation analysis technologies, using the expression

$$R^*_{X\varepsilon}(m) = \frac{1}{N} \sum_{i=1}^{N} \text{sgn} g(i\Delta t)[g(i+m)\Delta t + g((i+m+1)\Delta t - 2g(i+m+1)\Delta t)].$$

The obtained estimates allow determining the beginning of changes in the technical condition of rails. Similarly, using the corresponding formulas given in Chaps. 2–4, we calculate the estimates of the various noise characteristics reflecting the technical condition of the cars and rails.

During the training, over a certain time, the maximum threshold estimates of all noise characteristics are determined, at which the technical condition of the control object is considered normal. They form the reference set

$$W_j^{\max} = \begin{cases} D_{g_j\varepsilon}^{\max}, R_{g_jX\varepsilon}^{\max}(1\Delta t), R_{g_jX\varepsilon}^{\max}(2\Delta t), R_{g_jX\varepsilon}^{\max}(3\Delta t), \ldots, R_{g_jX\varepsilon}^{\max}(m\Delta t) \\ R_{g_jX\varepsilon}^{*\max}(1\Delta t), R_{g_jX\varepsilon}^{*\max}(2\Delta t), R_{g_jX\varepsilon}^{*\max}(3\Delta t), \ldots, R_{g_jX\varepsilon}^{*\max}(m\Delta t) \\ a_{g_jn}^{\max}, b_{g_jn}^{\max}; a_{g_jn\varepsilon}^{\max}, b_{g_jn\varepsilon}^{\max}; a_{g_jn\varepsilon_2}^{\max}, b_{g_jn\varepsilon_2}^{\max} \\ a_{g_jn}^{*\max}, b_{g_jn}^{*\max}; a_{g_jn\varepsilon}^{*\max}, b_{g_jn\varepsilon}^{*\max}; a_{g_jn\varepsilon_2}^{*\max}, b_{g_jn\varepsilon_2}^{*\max} \\ k_{g_jq_0}^{\max}, k_{g_jq_1}^{\max}, k_{g_jq_2}^{\max}, \ldots, k_{g_jq_m}^{\max}; \bar{f}_{g_jq_0}^{\max}, \bar{f}_{g_jq_1}^{\max}, \bar{f}_{g_jq_2}^{\max}, \ldots, \bar{f}_{g_jq_m}^{\max} \end{cases}.$$

Because of this, during the movement of the train, as a result of the use of various noise technologies to analyze vibration signals, the set of estimates is formed on the noise controller reflecting the current technical condition of cars, the entire train, rails, bridges, tunnels, and communications throughout the route.

If there is a defect, for instance, in the rail on which the car is currently moving, some current estimates of the noise characteristics will exceed the corresponding maximum threshold reference value. If there is a fault in the current state of the rail, the current estimates of some noise characteristics will be greater than the corresponding maximum reference estimates. The information about this will be transmitted via the interface of the train control system (TCS) for saving and displaying on the screen of the driver's monitor. The same will happen when the train crosses bridges and tunnels. The information received via radio channel from their noise controllers will reflect their technical condition. In addition, if the technical condition of the car changes, it will also affect the estimates of the specified noise characteristics of vibration signals. Naturally, all this information will be transmitted via the interface of the train monitoring system for registration and displaying on the screen of the driver's monitor.

Thus, throughout the route, the subsystem will provide noise control of the technical condition of individual cars and the whole train, as well as the railway tracks, bridges, tunnels, and communications. It is clear that by duplicating the relay correlation analysis of vibration signals using various noise analysis technologies, we enhance the reliability of the results of the use of the proposed system.

References

1. Aliev T (2003) Robust technology with analysis of interference in signal processing. Kluwer Academic/Plenum Publishers, New York. https://doi.org/10.1007/978-1-4615-0093-3
2. Aliev T (2007) Digital noise monitoring of defect origin. Springer, Boston. https://doi.org/10.1007/978-0-387-71754-8
3. Aliev TA, Guluyev GA, Pashayev FH, Sadygov AB (2012) Noise monitoring technology for objects in transition to the emergency state. Mech Syst Signal Process 27:755–762. https://doi.org/10.1016/j.ymssp.2011.09.005
4. Aliev TA (2016) Noise monitoring of accidents. Scholar's Press, Saarbrücken
5. Aliev TA, Alizada TA, Rzayeva NE (2017) Noise technologies and systems for monitoring the beginning of the latent period of accidents on fixed platforms. Mech Syst Signal Process 87:111–123. https://doi.org/10.1016/j.ymssp.2016.10.014
6. On-line diagnostic systems of WWER reactor. www.diaprom.com/en/projects/. Accessed 21 May 2018
7. Vibration noise diagnostic system of WWER-1000 reactor (VNDS). www.diaprom.com/en/projects/?p=1. Accessed 21 May 2018
8. Loose parts monitoring system in the main circulation circuit of WWER-1000 (LPMS). www.diaprom.com/en/projects/?p=2. Accessed 21 May 2018
9. Reactor Mode Diagnostic System of WWER-1000. www.diaprom.com/en/projects/?p=3. Accessed 21 May 2018
10. Online Monitoring for Improving Performance of Nuclear Power Plants. Part 1: Instrument Channel Monitoring. www.pub.iaea.org/books/IAEABooks/7790/On-line-Monitoring-for-Improving-Performance-of-Nuclear-Power-Plants-Part-1-Instrument-Channel-Monitoring. Accessed 21 May 2018
11. Online Monitoring for Improving Performance of Nuclear Power Plants, Part 2: Process and Component Condition Monitoring and Diagnostic. www.pub.iaea.org/books/IAEABooks/7908/On-line-Monitoring-for-Improving-Performance-of-Nuclear-Power-Plants-Part-2-Process-and-Component-Condition-Monitoring-and-Diagnostics. Accessed 21 May 2018
12. Advanced Surveillance, Diagnostic and Prognostic Techniques in Monitoring Structures, Systems and Components in Nuclear Power Plants. www.pub.iaea.org/books/IAEABooks/8763/Advanced-Surveillance-Diagnostic-and-Prognostic-Techniques-in-Monitoring-Structures-Systems-and-Components-in-Nuclear-Power-Plants. Accessed 21 May 2018
13. Accident Monitoring Systems for Nuclear Power Plants. www.pub.iaea.org/books/IAEAbooks/10754/Accident-Monitoring-Systems-for-Nuclear-Power-Plants. Accessed 21 May 2018
14. Instrumentation and Control Systems and Software Important to Safety for Research Reactors. www.pub.iaea.org/books/IAEAbooks/10719/Instrumentation-and-Control-Systems-and-Software-Important-to-Safety-for-Research-Reactors. Accessed 21 May 2018
15. Moghaddam AK (2017) A review on the current methods of railway induced vibration attenuations. Int J Sci Eng Appl 6(04):123–128. https://doi.org/10.7753/IJSEA0604.1001
16. Bezin Y (2016) Railway turnout damage prediction and design implications. In: International conference on train/track interaction & wheel/rail interface, 20–22 June 2016, Hall of Railway Sciences (CARS), Beijing, China

Chapter 8
Using Noise Control Technologies and Systems in Construction and Seismology

Abstract One of the possible solutions to the problem of monitoring the beginning of the initiation of anomalous seismic processes by carrying out of noise analysis of seismic-acoustic signals received from earth's deep strata is considered. The diagram of the station for noise monitoring of the beginning of seismic processes is proposed in which bores of suspended oil wells (1500–5000 m) are used as communication channels to receive seismic information from earth's deep strata. Experiments on shallow wells (40–200 m) have shown that they also reliably receive seismic-acoustic information, although within a shorter radius of reception of seismic-acoustic signals. The chapter analyzes the results of the experiments with the intelligent seismic-acoustic system for identifying the area of the focus of an expected earthquake based on the network consisting of seismic-acoustic stations built on six deep and four shallow wells. It has been established that only with the application of the noise technology, the beginning of seismic processes is reliably detected based on the estimates of the correlation function between the useful seismic-acoustic signal and its noise. The combinations of seismic-acoustic stations vary depending on their location and direction. If seismic processes from one direction are recorded by 3–5 stations, then completely different combinations of stations respond to them from the opposite direction. Due to this, the system can be used by seismologists as a tool for identifying the area of the focus of an expected earthquake based on the combinations of stations that respond to seismic processes.

8.1 Technology and System of Noise Control of Anomalous Seismic Processes

It is known that widely used seismic stations nowadays allow registering the moment of the beginning of an earthquake, determining the coordinates of its focus and magnitude. Various methods and technologies of spectral analysis are used to analyze seismic signals obtained from seismic sensors. It is also known that many variants of short-term earthquake forecasting have been proposed over several decades. The most successful of them give predictions with a sufficiently high degree of probability. However, earthquake forecasting has to be conducted with absolute reliability and validity. Unfortunately, existing earthquake prediction systems do not meet these requirements and therefore they have not found wide practical application. For this reason, in recent years, a pessimistic point of view has come to be with regard the unreality of solving the problem of creating technologies and systems that make it possible to implement a reliable short-term earthquake forecasting.

This section of the book is devoted to one of possible solutions to the problem of monitoring the onset of the initiation of anomalous seismic processes (ASP) by analyzing seismic-acoustic signals received from deep strata of the earth.

The monitoring of the beginning of ASP that result in earthquakes entails two specific problems. When an ASP arises, both infra-low frequency seismic waves and seismic-acoustic waves with a frequency within the sound range form. Both types of wave do not reach the surface of the earth for a long time before ASP reaches the critical state, which is explained by the fact that the frequency characteristics of the upper strata of the earth do not allow seismic-acoustic waves to reach the surface of earth. Seismic waves, on the other hand, only become powerful enough when ASP is in its critical state—when an earthquake is occurring. It follows that solving the given problem first of all requires obtaining seismic-acoustic noise from the deep strata of earth, it being the primary carrier of information on the incipient earthquake.

Another important aspect of the problem in question is related to the necessity of developing a technology of analysis of seismic-acoustic noise. It is known that the existing conventional technologies for the analysis of measurement information only yield satisfactory results under classical conditions such as normalcy of the distribution law, stationarity, absence of correlation between the useful signal and the noise, etc. Those conditions are, however, violated in seismic-acoustic noises as ASP arise and form. For this reason, traditional technologies cannot provide the sufficient reliability and adequacy of the obtained results. Thus, the second key issue of the problem in question comes down to the development of a technology that takes into account the peculiarities of a heavily noisy seismic-acoustic signal in the period of ASP formation. Here, the analysis of noise in the seismic-acoustic signal as a carrier of useful diagnostic information is of prime importance.

It is known that in seismically active regions, the time T_0 of the normal seismic state between occasional ASP varies within the range of several weeks or months. The time period T_1 of the origin of ASP and formation can last for several hours. The time period of the critical state T_2, when the seismic waves reach the surface of earth and an earthquake occurs, is estimated in minutes, after which the new period of rest T_0 begins. It is therefore appropriate to reduce the problem of monitoring and short-term forecasting of earthquakes to the ensuring a reliable indication of the beginning of the period T_1, which is the latent initiation of ASP. The known existing systems and widely used seismic stations are designed for registering the start of the period T_2. Unfortunately their functions do not include a reliable and adequate monitoring of the start of the period T_1, which is one of the severe drawbacks of modern systems and tools of both the control and monitoring of seismic processes.

In view of the above, the necessity of creating a system for receiving seismic-acoustic information from the deep strata of the earth and developing the noise technology for the analysis of noises as carriers of useful information is obvious.

8.1 Technology and System of Noise Control of Anomalous Seismic Processes

We consider a possible way to solve these problems in the following paragraphs. Theoretical and experimental studies have demonstrated that when an ASP originates at the start of the time T_1, the estimates of the noise variance D_ε, the cross-correlation function $R_{X\varepsilon}(\mu = 0)$ between the useful signal $X(i\Delta t)$ and the noise $\varepsilon(i\Delta t)$ are the first to change. The reason is that the noise $\varepsilon(i\Delta t)$ emerges due to random external factors, which have no correlation with the useful signal, in the time T_0 in the normal state of the seismic processes. However, in the time T_1, during ASP initiation, the noise $\varepsilon(i\Delta t)$ emerges due to the influence of the seismic processes. For this reason, in the period of time T_1, a correlation appears between the useful signal $X(i\Delta t)$ and the noise $\varepsilon(i\Delta t)$, and the inequalities $R_{X\varepsilon}(\mu) \neq 0$ take place.

It is therefore obvious that the estimates D_ε, $R_{X\varepsilon}(\Delta t)$, $R_{X\varepsilon}(2\Delta t)$, $R_{X\varepsilon}(3\Delta t)$, ..., $R_{X\varepsilon}(m\Delta t)$ should be used as informative attributes.

It is clear that in the period of time T_0 during the monitoring of ASP initiation, the estimates of the cross-correlation function $R_{X\varepsilon}(\mu = m)$ between the useful signal and the noise of the acoustic signal received from the deep strata of the earth at different time shifts will be close to zero, due to the absence of correlation between $X(i\Delta t)$ and $\varepsilon(i\Delta t)$. It is also obvious that during ASP initiation in the time period T_1, the value of their estimates will increase sharply due to the emergence of correlation between $X(i\Delta t)$ and $\varepsilon(i\Delta t)$. Thus, they will be different from zero from the very start throughout the course of ASP, reflecting the earthquake preparation process. Considering the high requirements to reliability and validity of the obtained results of ASP monitoring, the process of analyzing seismic-acoustic noise should be parallelized using several technologies.

The diagram of the station of noise monitoring of ASP is given in Fig. 8.1. Its first and foremost difference from all other known prototypes is that the steel bores of suspended oil wells filled with water are used for receiving seismic information from the deep strata of earth. Block 1 is the equipment for receiving seismic-acoustic noise from the deep strata of earth, based on a hydrophone installed at the head of a well. Seismic-acoustic signals are analysed by means of the above-mentioned noise technologies, and the corresponding estimates are calculated in block 2. Block 3 is standard seismic equipment that allows registering and assessing the intensity of seismic vibrations. Block 4 and the server of the monitoring centre are for identifying the results of ASP monitoring at the RNM ASP station with the earthquakes registered at seismic stations of the seismological service.

At the initial stage, corresponding estimates of seismic-acoustic signals received from the hydrophones of block 1 are calculated in block 2. Those estimates are transmitted to the monitoring server, where they are saved. The indicated estimates are formed and saved both in block 2 and on the server during the long period of time T_0. The estimates of the seismic signals coming from ground seismic equipment 3 are also calculated during time T_0 and registered in the same manner. This process continues until the moment when the current estimates of the signals received from corresponding sensors will differ from previous ones by a value greater than the set threshold levels.

The corresponding information is sent to the server at the Monitoring Center. Thus, information on the beginning of anomalous seismic processes forms in the

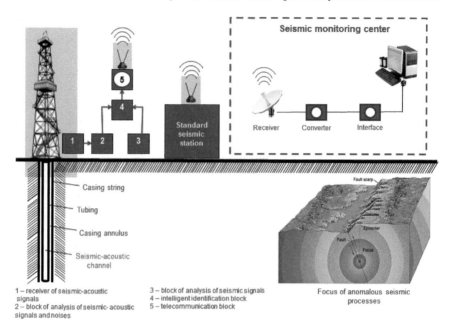

Fig. 8.1 Station of noise monitoring of ASP (RNM ASP): 1—receiver of seismic-acoustic signals; 2—block of analysis of seismic- acoustic signals and noises; 3—block of analysis of seismic signals; 4—intelligent identification block; 5—telecommunication block

system at the start of the time T_1 from the estimates of the seismic-acoustic signals received from the output of the hydrophones installed on the head of the steel well bore. Standard ground seismic stations, meanwhile, register corresponding signals and only determine the magnitudes of seismic vibrations only at the start of time T_2 (intense seismic vibrations). The information about this is also sent to block 4 and to the server, where the difference between the moments of receiving the corresponding signals in Blocks 2 and 3, respectively, is determined. Training and identification of ASP are carried out simultaneously both in block 4 and on the server during the operation, with the use of known technologies of recognition, including neural network technologies. After a certain training period on the server and in block 4, the time of registration of the moment of the expected earthquake by standard seismic stations is determined as a result of the monitoring of ASP.

Figure 8.2 shows the structure of the station—it is set up in a specially built room to protect the station from the sun, wind and other external factors.

The results of monitoring experiments at the RNM ASP station at Qum Island in the Caspian Sea are given below

An experimental version of RNM ASP station was installed at the head of a 3500 m deep suspended oil well #5 on 01.07.2010. The well is filled with water, and for this reason a BC 312 hydrophone is used as the sensor.

The following earthquakes have been registered by Azerbaijan seismic stations during the operation of the station from 01.07.2010 to 15.01.2011:

8.1 Technology and System of Noise Control of Anomalous Seismic Processes

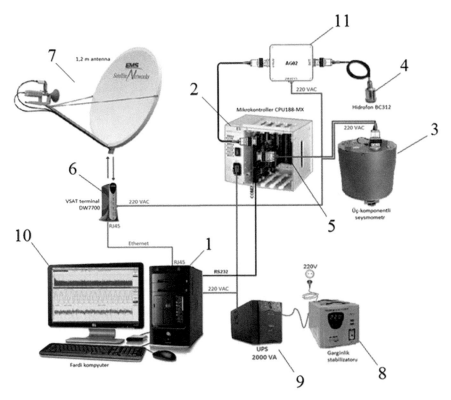

Fig. 8.2 Structure of RNM ASP station: 1 is system unit; 2 is Fastwell Micro PC controller; 3 is GURALP LTD CMG 5T seismic accelerometer; 4 is BC 321 hydrophone made in Zelenograd; 5 is reinforcing and normalising elements; 6 is Siemens MC35i terminal forming an Internet channel via GPRS; 7 is antenna; 8 is voltage regulator; 9 is UPS; 10 is monitor; 11 is connector

09.10.2010, town of Masally 00:58:11, M:3.5, d:12 km
11.10.2010, town of Shirvan, 22:50:23, M:3.9, d:37 km
17.10.2010, town of Imishli, 07:20:38, M:3.4, d:18 km
20.11.2010, Caspian Sea, 05:05:48-08:29:29, M:3.5, d: 50 km
25.11.2010, Baku, Sangachal, 09:15:21, M: 3.04, d: 36 km.

Given in Fig. 8.3a–e are the results of ASP monitoring by means of the RNM ASP. According to the records, the estimates between the useful signal and the noise of the seismic-acoustic signal received at from output of the hydrophone increase sharply more than 5–10 h before the earthquake. This continues until the end of the earthquake. It should be noted that the distance from an RNM ASP station to the remotest earthquakes is over 200–300 km. Figure 8.3f demonstrates the records of the estimate of the cross-correlation function $R_{X\varepsilon}(\mu)$ between the useful signal $g(i\Delta t)$ and the noise $\varepsilon(i\Delta t)$ related to the earthquakes in Azerbaijan (21.01.2011, 01:58:54), Georgia (23.01.2011, 07:51:23), Tajikistan (24.01.2011, 06:45:29) and on the border with Turkey, Armenia and Iran (3 earthquakes—25.01.2011, 03:56:12, 04:02:32,

07:40:04). As is clear from the results of the given charts of ASP monitoring, their lead over the beginning of the earthquake is over 5–10 h. For instance, Fig. 8.3g gives the expanded record of the estimate of cross-correlation function during the earthquake in Georgia (near Kutaisi) on 23.01.2011. It is obvious from the chart that the beginning of ASP 07:51:23 was clearly registered 5–6 h before the beginning of the earthquake. The charts show Baku time.

Experimental research has demonstrated that in the analysis of seismic-acoustic signals, clear identification of the beginning of the time T_1 by means of traditional technologies is impossible. The use of robust technologies of noise analysis of estimates $R_{X\varepsilon}(\Delta t)$ and D_ε, on the other hand, detect the beginning of initiation of anomalous seismic processes reliably and adequately.

Thus, the initial results of experiments show that it is possible to perform monitoring within a radius of over 200–300 km 5–15 h before the earthquake by means of RNM ASP. This means that the difference in time between ASP initiation and its critical state depends on the location of the earthquake focus. One can assume, based on the obtained results, that, when spreading from the earthquake focus, seismic-acoustic waves are reflected due to the resistance of certain upper strata of the earth and start propagating horizontally. One can also assume that sufficient intensity of those waves allows them to travel to long distances (300–500 km).

These experimental results show the expediency of building RNM ASP stations. With this in mind, eight more of such stations were built from 2012 to 2017, and the network of these stations made it possible to determine the directions and the focus areas of expected earthquakes.

In the long term, the integration of these seismic-acoustic stations with standard seismic stations will allow creating intelligent systems that, after a certain training period, will be capable of carrying out earthquake warning with sufficient practical degree of reliability and adequacy.

Thus, we can conclude from the obtained experimental results that the lead in the time of registration of ASP initiation by seismic-acoustic stations of RNM ASP over widely used standard seismic equipment is conditioned by two factors.

First, high-frequency seismic-acoustic waves, which arise at the start of initiation of anomalous seismic processes deep below the surface of earth, propagate through some strata horizontally in the form of noise that reaches the steel bore of the oil well at the depth of over 3–6 km. Serving as acoustic channels, the steel bores filled with water transmit the seismic-acoustic noise at the velocity of sound to the surface of earth, where it is received by means of the hydrophone in block 1. At the same time, infra-low-frequency seismic waves gain the required capacity in a certain amount of time, when seismic processes reach their critical state and an earthquake occurs, which is why they are registered by seismic receivers of standard ground equipment much later.

Second, the use of the noise technology allows analyzing noises as carriers of seismic information, which makes it possible to register anomalous seismic processes at the start of their initiation, and their detection by means of estimates of characteristics of useful signals starts considerably earlier. Thus, those two factors made it possible to detect the indication time of the start of ASP of the coming earthquake on

8.1 Technology and System of Noise Control of Anomalous Seismic Processes

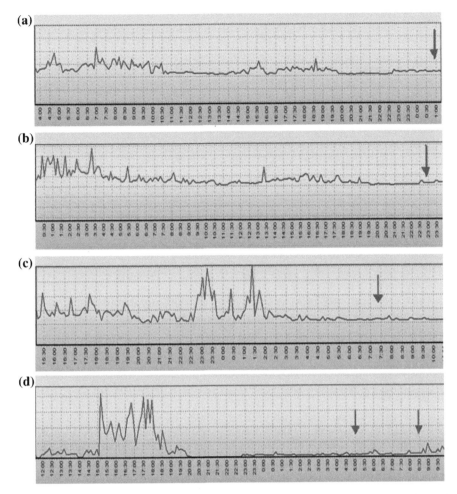

Fig. 8.3 **a** 08.10.2010, 08.10.2010 Masally, 00:58:11, M:3.5, d:12 km. Noise variance. Start of ASP approximately at 04:30, 08.10.2010, earthquake at 00:58:11, 09.10.2010. **b** 11.10.2010, Shirvan, 22:50:23, M:3.9, d:37 km. Noise variance. Start of ASP approximately at 00:30,11.10.2010, earthquake at 22:50:23,11.10.2010. **c** 16-17.10.2010, Imishli, 07:20:38, M:3.4, d:18 km. Noise variance. Start of ASP approximately at 15:30, 16.10.2010, earthquake at 07:20:38, 17.10.2010. **d** 19-20.11.2010. At sea, 05:05:48–08:29:29 20.11.2010, M:3.5, d: 50 km. Noise variance. Start of ASP approximately at 12:20, 19.11.2010, two earthquakes at 05:05:48, 08:29:29 20.11.2010. **e** 25.11.2010, Baku, Sangachal, 09:15:21, M: 3.04, d: 36 km. Noise variance. Start of ASP approximately at 12:10 24.11.2010, earthquake at 09:15:21 25.11.2010. **f** In Azerbaijan (21.01.2011, 01:58:54), in Georgia (23.01.2011, 07:51:23.0), in Tajikistan (24.01.2011, 06:45:29.0) and on the border between Turkey and Iran (3 earthquakes 25. 01.2011, 03:56:12.; 04·02·32.; 07:40:04.). **g** 23.01.2011, Georgia, near Kutaisi, 07:51:23, M: 4.5, d: 10 km. Estimates of cross-correlation function

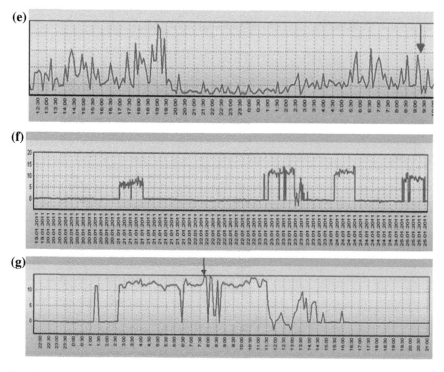

Fig. 8.3 (continued)

the basis of noise analysis of obtained seismic-acoustic data considerably earlier that it is registered by stations of seismological service. An earthquake can usually be detected no less than 5–15 h before its beginning, which can give a chance to warn the population in due time about the danger of a powerful earthquake. This potential of the stations was proved after the start of operational testing of the second station at the 4400 m deep well #427 in Shirvan Oil on 20 November 2011. The first experimental monitoring results given in Fig. 8.4 were obtained by this station on 23.10.2011 and 24.10.2011 more than 10–12 h before the beginning of the earthquake. The similar record was made at the Qum Island station.

Our research has demonstrated that to determine coordinates and magnitudes of expected earthquakes, we need to build networks consisting of at least 8–10 stations and to integrate them with standard seismic stations. To this end, another three stations were built in 2011 in addition to the station at Qum Island in the Caspian Sea: in the town of Shirvan to the south of the country, in the town of Siazan in the north, in the town of Naftalan in the west. Another four stations were built in the following years.

During the operation of all these seismic stations, the results of noise analysis of seismic-acoustic signals at the moment of the initiation of ASP are sent from each station to the server at the Monitoring Center. As a result, sets of informative attributes form.

8.1 Technology and System of Noise Control of Anomalous Seismic Processes

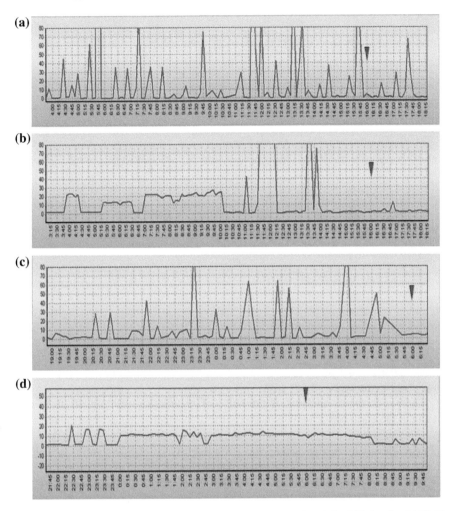

Fig. 8.4 **a** Shirvan 23 October 16:00:25 East Turkey M = 5.6. **b** Qum Island 23 October 16:00:25 East Turkey M = 5.6. **c** Shirvan 24 October 06:57:59 East Turkey M = 3.8 prolonged earthquakes. **d** Qum Island 24 October 06:57:59 East Turkey M = 3.8 prolonged earthquakes

After every ASP monitoring cycle, all stations perform system analysis on the server at the Monitoring Center based on the obtained sets of informative attributes.

According to the results of the experiments carried out at seismic-acoustic stations installed at the heads of 3–6 km deep oil wells in the period 01.05.2010–01.03.2012, traditional technologies for the analysis of seismic-acoustic signals received from hydrophones do not allow detecting the start of ASP initiation. The experiments have also demonstrated that the main carrier of information on the beginning of ASP preceding an earthquake is the noise of seismic-acoustic signals. The proposed technologies for the analysis of noise as a carrier of useful information calculate its

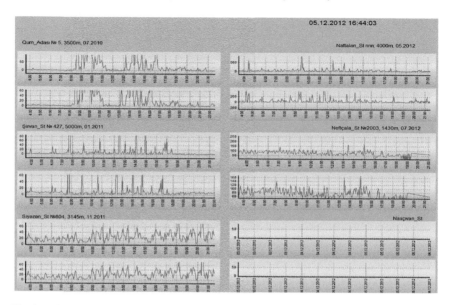

Fig. 8.5 Charts of the estimates of the seismic-acoustic noise obtained from the respective stations, as displayed on the screen of the Seismic-Acoustic Monitoring Center

characteristics, such as cross-correlation functions between the useful signal and the noise, noise variance.

Our synchronous analysis of the noise of seismic-acoustic signals from deep strata of the earth performed by means of the above technologies confirmed that the seismic-acoustic waves of ASP propagate within a radius of 300–500 km dozens of hours earlier than seismic waves registered by ground seismic stations, which is a prerequisite for their wide use in seismology.

All stations are connected to the Center of monitoring of anomalous seismic processes (CM ASP) via satellite Internet connection. On every station, the estimates of the characteristics of seismoacoustic noise are calculated with an interval of 5 s (by processing 5000 samples), from which packets are formed and transmitted to the CM ASP at an interval of 60 min. After the appropriate processing, the information is displayed on the server in the online mode and archived as well. The server provides for the possibility of displaying all archived files on the monitor and visual monitoring of the picture of ASP both in the Azerbaijan Republic and in neighboring regions. For instance, the monitor screen (Fig. 8.5) shows the charts of the changes in the estimates of the cross-correlation and variance of the noise of seismic signals received on 05.12.2012 from the RNM ASP stations. The charts of the estimates of the noise are arranged in the following sequence: Qum Island, Shirvan, Siazan, Neftchala, Naftalan, Nakhchivan.

As can be seen from the charts in Fig. 8.5, the registration of ASP at Naftalan station began at 04:30, at Shirvan station at 06:00, at Qum Island station at 08:30

and at Siyazan station at 10:00. The Neftchala station showed a slight change in the characteristics.

Taking into account the experience of observing and analyzing the experimental results of ASP registration over the past few months, it was concluded based on the records on the monitor that an earthquake was to be expected in the north-west of Azerbaijan. This was confirmed—there was an earthquake in the Caucasus region of Russia on 03.12.2012, at 18:58 Baku time. (GMT/UTC +4)

In conclusion, seismic-acoustic stations of ASP monitoring can also be used for monitoring the latent period of volcano formation well before the eruption. Their use will also allow one to perform the monitoring of testing of minor and major nuclear bombs and other experiments related to the manufacture of military equipment on a regional scale. A network of such stations will make it possible to fully control such tests and various military maneuvers.

8.2 Digital Citywide System of Noise Control of the Technical Condition of Socially Significant Objects

In countries located in seismically active zones, to ensure the safety of the population, regular monitoring of the technical condition of residential buildings and strategic facilities is required. The importance of this problem increases manifold for cases when, in addition to the seismic hazard, there is also a possibility of a landslide. Consequently, for cities located in such regions, it is relevant to create a citywide system for control of the technical condition of housing facilities and construction sites and signaling the beginning of anomalous seismic processes. Massive destructions that have disastrous consequences in these countries in recent years clearly show that the following problems must be solved:

1. Developing a digital technology and a system for regular monitoring of changes in the technical condition of socially significant construction facilities.
2. Developing a digital technology for regular monitoring of the technical condition of groups of facilities located in landslide zones.

The conducted studies have shown that if the estimates of the noise characteristics of signals received from vibration acoustic sensors installed on socially significant construction facilities located at significant distances from each other in seismically active regions change simultaneously, this indicates the beginning of anomalous seismic processes. If a change occurs in a single facility, this is due to the beginning of changes in the technical condition of the corresponding facility. Taking into account the specifics of the seismically active regions, Fig. 8.6 shows the diagram of one of the possible variants of the intelligent system of city-wide noise monitoring of construction facilities.

In this variant of the system, each of the facilities $O_1, O_2, \ldots, O_N$ is equipped with a local block with a "black box" based on a controller and acoustic, seismic

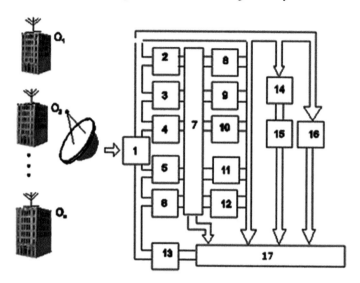

Fig. 8.6 Diagram of the noise monitoring system

and vibration sensors $D_1, D, \ldots, D_m$ installed in the most vulnerable elements of the structure. After the primary processing on the controller of the local system, the signals $g_1(i\Delta t), g_2(i\Delta t), \ldots, g_n(i\Delta t)$ are transmitted from each facility via communication channels to the Center of the noise monitoring system. Due to this, in the course of the operation of the system, block 1 receives signals from the monitored facilities that carry the information about facility's current technical condition.

During the operation of the intelligent noise control system, training is conducted at the first stage. For this purpose, blocks 2–6 designed for the noise analysis of the signals $g_1(i\Delta t), g_2(i\Delta t), g_3(i\Delta t), \ldots$ obtained from the corresponding sensors calculate the estimates of the noise variance D_ε, the cross-correlation function $R_{X\varepsilon}(1\Delta t)$, $R_{X\varepsilon}(2\Delta t)$ between the noise and the useful signal, the spectral characteristics $a_{n_\varepsilon}, b_{n_\varepsilon}$, etc. In the process of training, under facility's normal technical condition, in periods of time when low-power earthquakes occur, maximum values for of the estimates of the mentioned informative attributes are established. Due to this, the set of reference informative attributes is formed for the normal technical condition of each of the monitored facilities:

$$W_{g_j}^{\max} = \begin{cases} D_{g_j\varepsilon}^{\max}, R_{g_jX\varepsilon}^{\max}(1\Delta t), R_{g_jX\varepsilon}^{\max}(2\Delta t), R_{g_jX\varepsilon}^{\max}(3\Delta t), \ldots, R_{g_jX\varepsilon}^{\max}(m\Delta t) \\ R_{g_jX\varepsilon}^{*\max}(1\Delta t), R_{g_jX\varepsilon}^{*\max}(2\Delta t), R_{g_jX\varepsilon}^{*\max}(3\Delta t), \ldots, R_{g_jX\varepsilon}^{*\max}(m\Delta t) \\ a_{g_jn}^{\max}, b_{g_jn}^{\max}; a_{g_jn\varepsilon}^{\max}, b_{g_jn\varepsilon}^{\max}; a_{g_jn\varepsilon_2}^{\max}, b_{g_jn\varepsilon_2}^{\max}; \\ a_{g_jn}^{*\max}, b_{g_jn}^{*\max}; a_{g_jn\varepsilon}^{*\max}, b_{g_jn\varepsilon}^{*\max}; a_{g_jn\varepsilon_2}^{*\max}, b_{g_jn\varepsilon_2}^{*\max} \\ k_{g_jq_0}^{\max}, k_{g_jq_1}^{\max}, k_{g_jq_2}^{\max}, \ldots, k_{g_jq_m}^{\max}; \bar{f}_{g_jq_0}^{\max}, \bar{f}_{g_jq_1}^{\max}, \bar{f}_{g_jq_2}^{\max}, \ldots, \bar{f}_{g_jq_m}^{\max} \end{cases}$$

All of them are saved as reference informative attributes in blocks 8–12.

After the training stage, the system goes into monitoring mode. During earthquakes, current estimates of the noise variance, the correlation function between the useful signal and the noise, the spectral characteristics of the noise, etc. are calculated by blocks 2–6. They are compared with the corresponding reference values registered at the training stage in blocks 8–12. If the difference does not exceed the established minimum ranges, then the technical condition of the corresponding object $O_1, O_2,\ldots, O_N$ is considered unchanged. Otherwise, the resulting differences are sent to the decision-making block 17, where the information is generated about the beginning of changes in the technical condition of the corresponding facility. The value of the difference in the deviation range determines the severity of the situation.

In the second mode, unlike the first one, only if a simultaneous deviation above the maximum range of the estimates obtained from closely located groups of objects is detected, a signal about the beginning of a landslide is generated.

The third mode differs from the second one in that the signaling about the beginning of anomalous seismic processes is generated if a deviation above the maximum range of the estimates from several groups of objects located at considerable distances from each other is detected simultaneously. In block 17, the severity of the seismic situation is assessed based on the number of groups, especially the value of the deviation. For cases where deviation values exceed the maximum threshold levels, block 17 generates a signal about the threatening nature of the beginning of anomalous processes.

In conclusion, we should point out several factors that show the importance of developing and creating the described system.

1. When traditional technologies of control of the technical condition of a construction facility are used, a change in it is detected only at the time when the change takes on a pronounced form. Meanwhile, the use of the noise analysis technology for this purpose makes it possible to detect such a change at the initial stage. Because of this, through early detection of minor defects, it is possible to organize timely preventive measures and prevent the occurrence of serious defects, which allows significantly reducing repair costs and the amount of sudden destruction.
2. The proposed city-wide noise monitoring system allows detecting the initial stages of a landslide when there occurs a simultaneous change in the ranges of deviations of the corresponding estimates of the noise of the signals received from a group of facilities located in a certain area of the city. Due to this, it is possible to inform the competent city services of the possibility of a destructive and environmentally hazardous process in advance.
3. The relevance and expediency of creating a citywide noise monitoring system is largely due to the fact that if the ranges of deviations of the values of the corresponding estimates of the noise of the signals received from sensors of facilities located in different parts of the city exceed the established threshold values, then it is possible to reliably detect the initial stages of anomalous seismic processes and make decisions on appropriate signaling.

8.3 Intelligent Seismic-Acoustic System for Identifying the Area of the Focus of an Expected Earthquake

It is known that there have been a lot of intensive studies in the recent years devoted to obtaining the information about anomalous seismic processes that precede powerful earthquakes [1–7]. Seismic signals received during earthquakes are analyzed by means of wavelet transform and finite elements [4, 8–11]. Earthquake prediction-related problems remain as a primary trend of research [12–16]. Various models and tools have been and are being developed; many earthquake early warning systems for general population models and technologies for prompt response of rescue groups of relevant authorities have been developed and introduced [17–22]. Despite the above-mentioned works, earthquakes are still not predicted in due time, which leads to many catastrophic consequences.

The operating principle of the seismic-acoustic system for monitoring the beginning of the earthquake preparation process is described in the previous secion. The system consists of the network of nine seismic-acoustic stations of robust noise monitoring of anomalous seismic processes (RNM ASP). Our experiments on those stations carried out from 2010.07.01 have established that a cross-correlation between the noise and the useful signal of seismic acoustic data appears during ASP initiation.

The results of operation of these stations has demonstrated that, using the varying estimate of the cross-correlation function between the useful signal and the noise, each of the stations in the network separately is reliable enough to indicate ASP initiation processes preceding earthquakes. However, the accuracy of coordinates of a coming earthquake determined by means of those stations proves to be insufficient. However, it has been established by way of experiments that it is possible to create an intelligent neural network system that would use these stations for locating the ASP area. The following is one of the possible way to build said system.

To solve this problem, we shall first consider the possibility of using for this purpose existing methods for calculating the earthquake epicenter on the basis of the seismic information obtained through the network of standard ground seismic stations [23, 24]. In these cases, the difference in the time it takes of the main seismic waves P and S to reach the ground stations is used to determine the epicenter of earthquakes. The propagation velocity of P waves is higher than that of S waves. In a homogeneous isotropic medium, the wave velocity P is calculated from the expression.

$$V_P = \sqrt{\frac{k + \frac{4}{3}\mu}{\rho}},$$

where k is the volume factor, μ is the shear modulus and ρ the density of the medium that waves penetrate.

The velocity of S wave propagation can be calculated from the expression

8.3 Intelligent Seismic-Acoustic System for Identifying the ...

$$V_S = \sqrt{\frac{\mu}{\rho}},$$

where μ is the shear modulus and ρ is the density of the material penetrated by the waves.

The distance from a regular ground seismic station to the epicenter is calculated by multiplying the time difference by the difference in velocity:

$$S = \Delta T(v_p - v_s).$$

After the distance between the epicenter and the different seismic stations has been determined, the coordinates of the focus are found geometrically. Unfortunately, in all known cases, the coordinates of epicenters and hypocenters in seismic monitoring systems are determined after actual earthquakes.

Our experimental research showed that, for many reasons, it is practically impossible to use the results obtained on RNM ASP stations to calculate the coordinates of the ASP areas by means of said technology.

In view of the above, considered below is one of the possible ways to create an intelligent system based on the networks of seismic-acoustic stations both for monitoring ASP initiation and locating the location of its area, and for determining the approximate value of the magnitude of an expected earthquake.

As was mentioned earlier, during ASP initiation at the start of the time T_1 the first estimates to change are those of cross-correlation function $R_{X\varepsilon}(\mu = 0)$ between the useful signal $X(i\Delta t)$ and the noise $\varepsilon(i\Delta t)$, and the noise variance D_ε [25–35]. This happens because the noise $\varepsilon(i\Delta t)$ forms due to the effects of ASP at the start of time T_1. Consequently, a correlation emerges between the useful signal $X(i\Delta t)$ and the noise $\varepsilon(i\Delta t)$, causing an abrupt increase in the estimate $R_{X\varepsilon}(\mu)$ in the period T_1. Therefore, we can consider $R_{X\varepsilon}(\mu)$ the main informative attribute that should be used in the monitoring of the latent period of ASP initiation.

Starting from 2010.07.01, we used traditional technologies as well as noise technologies on RNM ASP stations to register the start of the latent period of ASP initiation. A sufficiently reliable registration of the period T_1 by means of the estimates obtained through traditional spectral and correlation technologies proved to be unattainable. The use of the noise technology, however, caused an abrupt change in the estimates of the cross-correlation function $R_{X\varepsilon}(\mu)$ and the value of the noise variance D_ε at the start of the time T_1. This became a crucial factor that allows performing the monitoring of the beginning of ASP initiation with a sufficient degree of reliability. For this reason, creating the network of RNM ASP stations, we chose the estimates of $R_{X\varepsilon}(\mu)$ and D_ε as informative attributes in solving the problem of monitoring of the start of ASP initiation.

Our experimental research have demonstrated that the estimates of the relay cross-correlation $R^*_{X\varepsilon}(\mu = 0)$ between the useful signal $X(i\Delta t)$ and its noise $\varepsilon(i\Delta t)$ should be used as reliable informative attributes. Those estimates can be calculated from the expression

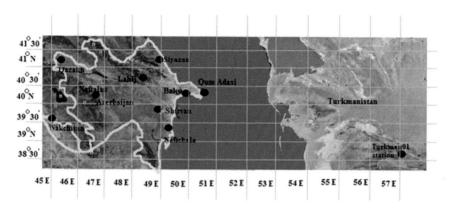

Fig. 8.7 Map of the locations of RNM ASP stations in the seismically active Caspian region

$$R^*_{X\varepsilon}(\mu = 0) \approx \frac{1}{N} \sum_{i=1}^{N} \Big[\text{sgn}g(i\Delta t)g(i\Delta t) - 2\text{sgn}g(i\Delta t)g((i+1)\Delta t)$$
$$+ \text{sgn}g(i\Delta t)g((i+2)\Delta t)\Big].$$

The experiments have shown that to improve the reliability and validity of ASP monitoring results, we should also use the estimates $R_{X\varepsilon}(1\Delta t)$, $R_{X\varepsilon}(2\Delta t)$, $R_{X\varepsilon}(3\Delta t)$ and the noise variance D_ε as additional informative attributes.

We shall now proceed to consider the results of our experimental studies using the networks of RNM ASP stations.

It is known that an earthquake takes place as soon as an ASP achieve its critical state. The boundaries of the area and the magnitude of an earthquake depend on the structure and nature of the strain–stress distribution in the rocks in a particular location. A jump-like rock deformation emits elastic waves. The amount of the deformed rocks determines the intensity of the shock and the formation of the seismic-acoustic noise $g(i\Delta t)$. Every core burst is preceded by quite a long period T_1 of earthquake preparation that can last up to dozens of hours.

Analyzing the seismic data obtained by means of acoustic sensors installed at the head of suspended oil wells, we find that during ASP initiation, seismic-acoustic noise traveling in the earth's deep layers anticipates an expected earthquake by dozens of hours T_1. Experiments have established that RNM ASP stations can quite reliably monitor the beginning of the time T_1 by the above-described technology [25–35].

In the following paragraphs, we consider developing an intelligent technology for locating the ASP area, using the data from the stations installed in nine seismically active regions of the Caspian Sea (Fig. 8.7). The geographical coordinates and well depths of the stations are given in Table 8.1.

Our experiments on the RNM ASP stations (Fig. 8.7) have shown that the seismic-acoustic noises received by hydrophones installed at the head of wells from the earth's deep layers are immediate precursors of the earthquake preparation process.

8.3 Intelligent Seismic-Acoustic System for Identifying the … 161

Table 8.1 Well depths and geographical coordinates of RNM ASP stations

Qum Island	40.310425°	50.008392°	3500 m	Jul	2010
Siazan	41.046217°	49.172058°	3145 m	Nov	2011
Naftalan	40.609521°	46.791458°	4000 m	May	2012
Shirvan	39.933170°	48.920745°	4900 m	Nov	2011
Neftchala	39.358333°	49.246667°	1430 m	Jun	2012
Nakhchivan	39.718000°	44.876000°	1800 m	Mar	2013
Qazakh	41.311889°	45.100278°	200 m	Aug	2013
Turkmenistan	38.530089°	56.654472°	300 m	Aug	2013
Baku (Cybernetics)	40.375700°	49.810833°	10 m	Feb	2014

The results of the measurement and analysis of those noises are transmitted from every station to the server of the Center for Seismic-acoustic Monitoring (CM) via satellite communication. The system also has the feature of forwarding the obtained results to the serves of CMs in neighboring countries in seismically active regions.

Starting from 01.07.2010, RNM ASP stations Qum Island (Caspian Sea), Shirvan, Siazan, Naftalan, Neftchala, Nakhchivan (on the borders with Iran and Turkey), Kopetdag (Turkmenistan), Qazakh (on the border with Georgia), Cybernetic (Baku) were put into operation one by one to conduct large-scale ASP monitoring experiments. The last three stations were built on 300 m, 200 m, and 10 m deep water wells, respectively. Those wells are filled with water by gravity flow. Hydrophones are placed in the water at the depth of 10-20 m from the water level. Our analysis of the seismic-acoustic signals received by these stations shows that during ASP initiation seismic-acoustic noises emerge spreading tens of hours earlier than seismic waves that are registered by the ground stations during an earthquake.

A synchronous robust noise analysis of the seismic-acoustic signals received from all stations via radio communication channels is performed during the operation of the network. The estimates of the noise characteristics $R_{X\varepsilon}(1\Delta t)$, $R_{X\varepsilon\varepsilon}(2\Delta t)$ and D_ε are sent to the server of the CM from the stations every 5 s. Based on the changes in those estimates, the identification of the starting points T_{1i} and T_{1j} of ASP initiation for the i-th and j-th stations, respectively, is carried out.

The result of the operation of these stations has shown that each of them individually makes it possible to reliably perform the indication of the process of initiation of ASP preceding an earthquake. It also became clear that we can use the results obtained by means of the network of these stations to create an intelligent technology for identifying the location of the area of an anticipated earthquake. For this purpose, by means of the network, we first determine the combinations of indication moments T_{1i} and T_{1j}, which together with the location coordinates of the stations represent the initial data for the solution of the problem of locating the ASP area. For the results to be more adequate and trustworthy, it is appropriate to use, in addition to the combinations of indication moments, time differences $T_{1i} - T_{1j}$ for each chosen pair of stations. In other words, to solve the problem at hand, we should determine not only the combination T_{1i}, T_{1j} but also the difference in time of ASP indication between the stations, i.e. the differences $\Delta\tau_{ij} = (T_{1i} - T_{1j})$.

According to the results of our experiments, it is not easy to determine the start of the time of indication T_{1i} with sufficient accuracy by means of the estimates of noise characteristics. For this reason, taking into account the importance and necessity of improving its accuracy, the proposed system duplicates the process of determining $\Delta\tau_{ij}$. To this end, it was found expedient to also determine the time difference $\Delta\tau_{ij} = (T_{1i} - T_{1j})$ using the extreme value of the estimate of the cross-correlation function $R_{g_i g_j}(\mu_{max})$ between the signals $g_i(i\Delta t)$ and $g_j(i\Delta t)$ obtained from different combinations of RNM ASP stations. The following expressions are used for this purpose:

$$R_{g_i g_j}(\mu_{max}) = \frac{1}{N} \sum_{i=1}^{N} g_i(i\Delta t) g_j(i+\mu)\Delta t,$$

$$R^*_{g_i g_j}(\mu_{max}) = \frac{1}{N} \sum_{i=1}^{N} g_i^2(i\Delta t) g_j^2(i+\mu)\Delta t,$$

$$R^{**}_{g_i g_j}(\mu_{max}) = \frac{1}{N} \sum_{i=1}^{N} g_i(i\Delta t) g_j^2(i\Delta t).$$

In this case, the procedure for determining the difference in monitoring time between different stations on the CM server comes down to the following:

1. finding the time of registration of the start of the period T_{1i} of ASP initiation by the first station (Qum Island);
2. finding the time of registration for the second (Shirvan), third (Siazan), fourth (Naftalan) station, etc.;
3. calculating the sets of estimates of the cross-correlation functions $R_{g_i g_j}(i\Delta t)$, $R^*_{g_i g_j}(i\Delta t)$, $R^{**}_{g_i g_j}(i\Delta t)$ from the corresponding expressions, choosing from the results the time shifts $\mu \Delta t$, at which the curve of the cross-correlation function has the peak (extreme) value and using those time shifts to calculate $\Delta\tau_{ij} = (T_{1i}-T_{1j})$, i.e. the difference in ASP registration time by the i-the and the j-the station, respectively;
4. the found time differences $\Delta\tau_{1i} = (T_{1i} - T_{1j})$ are used as the source data to identify the location of ASP area.

Thus, in the proposed system, the estimates of the noise characteristics $R_{X\varepsilon}(1\Delta t)$, $R_{X\varepsilon\varepsilon}(2\Delta t)$ and D_ε obtained as a result of ASP monitoring by RNM ASP stations Qum Island, Shirvan, Siazan, Naftalan, Neftchala and Nakhichevan, Turkmen01 (Turkmenistan), Qazakh and Cybernetic (Baku) are synchronously transmitted via the communication channel to the server of the CM. On the basis of the obtained monitoring results, combinations of sequences of indication times T_{1i}, T_{1j} and combinations of time differences $\Delta\tau_{ij}$ are formed and used as the source data for identifying the location of the area of an expected earthquake.

The experiments on the said stations carried out in the period 2010.07.01 to 2014.06.01 have demonstrated that there are the following active earthquake areas in Azerbaijan and neighboring regions within a radius of 300–500 km around the network of RNM ASP stations:

I. Turkmen coast of the Caspian Sea;
II. south of the Absheron peninsula (in the Caspian Sea);
III. north of the Absheron peninsula (in the Caspian Sea);
IV. Shirvan (region of Azerbaijan);
V. north-western regions of Azerbaijan;
VI. southern regions of Azerbaijan;
VII. south of the Caucasus region of the Russian Federation;
VIII. north-eastern regions of Iran;
IX. north-western regions of Iran (near Tabriz);
X. Iranian–Iraqi–Turkish border;
XI. northern regions of Iran;
XII. eastern regions of Turkey;
XIII. western regions of Georgia (Black Sea).

Some of the results ASP registration by RNM ASP stations in those areas are given in [28, 29].

Many earthquakes with magnitude 3–4 occurred in these areas. Combinations of the sequence of the times of ASP indication by Qum Island, Shirvan, Siazan, Neftchala, Naftalan and Nakhchivan stations practically repeated themselves. Our analysis of the recorded charts has demonstrated that each sequence combination of time of indication of current ASP corresponds to one of the concrete earthquake areas. The experience of their operation also demonstrates that the results obtained by means of these stations open the way for creating an intelligent technology to identify the areas of expected earthquakes. It became obvious that the network of these stations could be used as a tool for identifying the areas of expected earthquakes After studying the problem of interpretation of the experimental materials for over 2 years, we have learned to identify the location of the area of an expected earthquake, error-free intuitively, using these combinations of time. It then became obvious that the problem of identifying the location of an expected earthquake should be solved by using expert systems (ESs). This, in its turn, demonstrated the possibility of creating an ES which will allow seismologists to use the network of the RNM ASP stations as a toolkit to determine the location of the area of an expected earthquake.

The experimental version of the expert system for identifying the location of the ASP area (ESILA) proposed in this paper is based on the knowledge base (KB) comprising the totality of the sets W_1, W_2, W_3, ..., W_{13} of the locations of the respective areas. The elements of each of these sets are formed from the data of the charts recording the parameters of all earthquakes registered by the RNM ASP stations in all 13 areas from 2010.07.01 to the present day. Each element consists of the combination of the sequence of times of ASP indication by the stations T_{1i}, T_{1j}, the combination of the differences in times of indication $\Delta \tau_{ij}$, and the combination of the estimates of the cross-correlation function $R_{X\varepsilon}(\mu = 0)$. Each element of the

KB also contains the value of magnitude M_i determined during the corresponding earthquakes by ground seismic stations. The date of the earthquake is also indicated in each element. In the case when there is only one element, the KB looks as follows:

$$W_1 = \begin{Bmatrix} T_{11}^1 & T_{12}^1 & \ldots & T_{19}^1 & M_1 \\ \Delta\tau_{12}^1 & \Delta\tau_{13}^1 & \ldots & \Delta\tau_{19}^1 & M_1 \\ R_{1X\varepsilon}^1(\mu=0) & R_{2X\varepsilon}^1(\mu=0) & \ldots & R_{9X\varepsilon}^1(\mu=0) & M_1 \end{Bmatrix},$$

$$W_2 = \begin{Bmatrix} T_{11}^2 & T_{12}^2 & \ldots & T_{19}^2 & M_2 \\ \Delta\tau_{12}^2 & \Delta\tau_{13}^2 & \ldots & \Delta\tau_{19}^2 & M_2 \\ R_{1X\varepsilon}^2(\mu=0) & R_{2X\varepsilon}^2(\mu=0) & \ldots & R_{9X\varepsilon}^2(\mu=0) & M_2 \end{Bmatrix},$$

$$W_{13} = \begin{Bmatrix} T_{11}^{13} & T_{12}^{13} & \ldots & T_{19}^{13} & M_{13} \\ \Delta\tau_{12}^{13} & \Delta\tau_{13}^{13} & \ldots & \Delta\tau_{19}^{13} & M_{13} \\ R_{1X\varepsilon}^{13}(\mu=0) & R_{2X\varepsilon}^{13}(\mu=0) & \ldots & R_{9X\varepsilon}^{13}(\mu=0) & M_{13} \end{Bmatrix}.$$

Each set $W_1 - W_{13}$ of the experimental version of the KB consists of several dozens of such elements and new elements are added during each new earthquake. During the operation of the ES, after the monitoring and indication of the time of the beginning of the current ASP has been completed, the stations form current combinations of the sequence of times of indication T_{1i}, T_{1j} the combinations of differences in times of indication $\Delta\tau_{1j}$, as well as the combinations of the values of the estimates of $R_{X\varepsilon}(\mu)$.

On 2014.01.05, the experimental phase of the identification of the location of earthquake areas by ESILA was launched. This is carried out as follows. The current element is formed on the basis of the results of monitoring carried out by the network of RNM ASP stations. After that, the current element is compared with all elements of the sets W_1, W_2, W_3,..., W_{13} in the identification unit of the expert system (IBES). If it matches any element of any set, the location of the area of an expected earthquake is identified based on the number of that set. The number of the ASP area is saved in the decision-making block (DMB) of the ES. At the same time, the current element is entered into the set of the KB. Thus, new elements are continuously written into the KB during the ESILA operation and the network of RNM ASP stations and ESILA operate as a single system.

To check the validity and reliability of the identification of the location of the ASP area, the described ESILA was tested during all subsequent earthquakes. The obtained results have demonstrated the real possibility of practical application of the experimental version of ESILA to identify the location of the ASP area, which creates prerequisites for using the system in question as a toolkit for determining the location of the area of an expected earthquake. Taking this prospect into account, a function of forming the following information and providing it to seismologists was included in the list of basic functions of the DMB of ESILA:

1. Date of the current ASP and the number of the area of the expected earthquake.
2. Results of the current monitoring performed by the RNM ASP stations.

8.3 Intelligent Seismic-Acoustic System for Identifying the ...

3. Estimated lead time at the beginning of ASP monitoring compared with the time of registration of the expected earthquake by ground seismic stations.
4. List of all elements previously registered in the corresponding set during the initiation of the previous ASP in the estimated location of the area of the expected earthquake (with dates).
5. Amount of elements matching the current elements.
6. Magnitudes of previous earthquakes.
7. Minimum magnitude of the expected earthquake.
8. If the KB contains no elements matching at least some of the elements in the sets W_1–W_{13}, information on the impossibility of identifying the earthquake area is formed in the DMB.

The analysis of the results of the experimental identification of the location of the ASP area has demonstrated that, knowing the current estimates $R_{X\varepsilon}(\mu)$, $R_{X\varepsilon\varepsilon}(\mu)$, D_ε and the distance from the area to the RNM ASP stations, it is possible to calculate the approximate value of the minimum magnitude of an expected earthquake using neural networks [28, 29]. It was found that it is appropriate to use the information contained in the sets W_1–W_{13} to train neural networks, which then will function in the following way. The content of the corresponding elements of the sets W_1, W_2, W_3,..., W_{13} is transmitted to the outputs X_1, X_2,..., X_{N1} of the neuron, i.e. the combinations of times of the ASP indication T_{1ij}, differences of indication time $\Delta\tau_{ij}$ and the estimate $R_{X\varepsilon}(\mu = 0)$ are received at the inputs of the neuron one by one; the magnitude M_1 of the earthquake registered by ground stations is established at the output of the neuron. The training is carried out successively from earthquake area I to earthquake area XIII. For instance, during the training of the neuron on area III, i.e. during the earthquake with the area in the Caspian Sea, the monitoring results obtained at the stations in Siazan, Qum Island, Neftchala and Turkmen01 (Turkmenistan) are successively transmitted from the KB to the inputs of the neuron. The value of the magnitude M_3 is obtained at the output. During the training of the neuron to determine the magnitude in area XII, i.e. in East Turkey, then the monitoring data of Qazakh, Naftalan, Shirvan and Nakhchivan stations are sent to the input of the neuron and the magnitude M_{12} goes to the output. Thus, the parameters of ASP monitoring previously registered by the RNM ASP stations are used for the neural network training. At the same time, the coordinates of the location of the earthquake area are used in the DMB to determine the approximate distance S_I between the stations and the areas, which are also transmitted to the inputs of the neural network. Based on the source data written in the elements of the sets W_1–W_{13} and the distances $S_1 - S_9$ from the ASP area to each station, the neural network learns to determine the approximate magnitude of an expected earthquake. Due to this, after the training stage and in the process of current ASP monitoring, when the current combinations of corresponding estimates are transmitted to the neuron outputs, the code of the corresponding magnitude M of the expected earthquake forms at the output y_3 [28, 29]. The result is sent to the input of the DMB of ESILA.

During the operation of the neural network and the expert system, every time the coordinates and approximate magnitude of every expected earthquake have been identified, the obtained results are compared with the coordinates and magnitude of actual earthquakes registered by ground seismic stations. The resulting difference is further used to correct the KB and in the training of the neural network. Due to this, the KB is improved in the course of time, with the training level of the neural network constantly improving as well. This results in increased reliability, validity and adequacy of identification of the location and magnitude of expected earthquakes.

An analysis of our experience of the use of the ES in identifying the location of the area of the expected earthquake and of the neural network in determining its magnitude has demonstrated that improved reliability and authenticity of the obtained results requires increasing the number of RNM ASP stations. With this in mind, since the beginning of 2013, a station in the Nakhchivan Autonomous Republic near the border with Turkey and Iran and Turkmen01 station in Turkmenistan have been commissioned. On 2013.07.01, another station was built in the Qazakh region on the border with Georgia and at the Institute of Cybernetics, Baku (see Fig. 8.7).

It should be noted that the results of the experimental monitoring of the station built in the basement of the Institute of Cybernetics on a 10 m deep well practically match the readings of the Qum Island station built in a 3500 m deep well.

As was mentioned earlier, the test operation of the system under discussion started on 2014.01.06. In this period, some identification errors were detected during earthquakes with magnitude below 2.5–3.5. Besides, we also detected erroneous identification results at 2–3 RNM ASP stations simultaneously during a malfunction of the power supply system, communication systems and hydrophone, controller and other units. In the normal state of operation of all RNM ASP stations, no errors were detected in the results of the identification of the location of areas of expected earthquakes stronger than 5 points.

The list of the location of areas of expected earthquakes identified from the archived monitoring results in 2013–2014 is very long, which is why Table 8.2 contains only the results of 11 identified areas of expected earthquakes with magnitude over 5 from 2013.01.01 to 2014.07.06. Figure 8.8a–j are the recorded ASP charts that preceded those earthquakes. The data in the first column in Table 8.2 are taken from the website of the Euro-Mediterranean Seismological Centre (EMSC) (www.emsc-csem.org/#2).

The time of the earthquakes in Table 8.2 is given in UTC as provided by the EMSC website. The charts show the local time (Baku time), which is UTC+4 in winter and UTC+5 in summer.

Column 22 of Table 8.2 provides the information on the locations of the identified areas of expected earthquakes. To demonstrate the validity of those results, each row of the Table 8.2 is accompanied by relevant charts (Fig. 8.8a–j) recorded by the RNM ASP stations during the initiation of the respective ASPs.

The magnitudes of the earthquakes are given in units ML, mb, Mw in column 2 of the Table 8.2.

Sign "−" in Table 8.2 means that the value of the registered value of $R_{X\varepsilon}(\mu)$ was lower than the threshold level.

8.3 Intelligent Seismic-Acoustic System for Identifying the ... 167

Table 8.2 Identified areas of expected earthquakes

#	Date, time, coordinates, magnitudes and depth of earthquake epicenter	$\Delta\tau_{12}$	$\Delta\tau_{13}$	$\Delta\tau_{14}$	$\Delta\tau_{15}$	$\Delta\tau_{16}$	$\Delta\tau_{17}$	$\Delta\tau_{18}$	$\Delta\tau_{19}$	$R_{IX\varepsilon}$
1	2	4	5	6	7	8	9	10	11	12
1	2013-03-26 23:35:24.0 UTC 43.219 N; 41.637 E mb 5.1; 10 km	35	−120	−	135	−	−	−	−	300
2	2013-05-28 00:09:52.0 UTC 43.22 N; 41.58 E mb 5.2; 2 km	−115	−150	−250	−	−	−	−	−	150
3	2013-09-17 04:09:14.0 UTC 42.17 N; 45.89 E mb 5.0; 10 km	−	−150	−	390	−	120	−	−	100
4	2013-09-17 04:09:13.0 UTC 42.13 N; 45.80 E Mw 5.1; 2 km	−	−150	−	390	−	120	−	−	100
5	2013-11-24 18:05:41.0 UTC 34.06 N; 45.52 E mb 5.6; 2 km	−	−	−	−10	−60	−	−	−	160
6	2014-01-10 00:45:32.0 UTC 41.77 N; 49.31 E ML 5.0; 87 km	−	20	−	110	−	−	−10	−	110

(continued)

Table 8.2 (continued)

#	Date, time, coordinates, magnitudes and depth of earthquake epicenter	$\Delta\tau_{12}$	$\Delta\tau_{13}$	$\Delta\tau_{14}$	$\Delta\tau_{15}$	$\Delta\tau_{16}$	$\Delta\tau_{17}$	$\Delta\tau_{18}$	$\Delta\tau_{19}$	$R_{IX\varepsilon}$
1	2	4	5	6	7	8	9	10	11	12
7	2014-01-14 13:55:02.0 UTC 40.38 N; 52.97 E mb 5.1; 50 km	–	−45	–	−120	–	–	–	–	160
8	2014-01-28 23:47:38.0 UTC 32.52 N; 49.98 E ML 5.1; 33 km	−135	–	–	100	−300	–	–	–	120
9	2014-02-10 12:06:48.0 UTC 40.23 N; 48.63 E Mw 5.4; 55 km	−300	–	–	–	45	75	–	–	75
10	2014-06-07 06:05:32.1 UTC 40.32 N; 51.55 E mb 5.6; 50 km	145	20	–	−70	–	–	–	120	80
11	2014-06-29 17:26:10.4 UTC 41.62 N; 46.68 E mb 5.1; 20 km	305	−85	–	–	–	315	–	–	100

(continued)

8.3 Intelligent Seismic-Acoustic System for Identifying the ...

Table 8.2 (continued)

#	$R_{2X\varepsilon}$	$R_{3X\varepsilon}$	$R_{4X\varepsilon}$	$R_{5X\varepsilon}$	$R_{6X\varepsilon}$	$R_{7X\varepsilon}$	$R_{8X\varepsilon}$	$R_{9X\varepsilon}$	Number and location of the area of expected earthquake
13	14	15	16	17	18	19	20	21	22
1	50	100	–	140	–	–	–	–	Georgia (Sak'art'velo)
2	150	160	250	–	–	–	–	–	Georgia (Sak'art'velo)
3	–	40	–	80	–	80	–	–	Caucasus Region, Russia
4	–	40	–	80	–	80	–	–	Caucasus Region, Russia
5	–	–	–	150	250	–	–	–	Iran-Iraq Border
6	–	110	–	110	–	–	40	–	Caspian Sea, Offshore Azerbaijan
7	–	100	–	120	–	–	–	–	Turkmenistan
8	110	–	–	–	180	–	–	–	Western Iran
9	130	–	–	–	260	230	–	–	Azerbaijan
10	25	40	–	100	–	–	–	80	Offshore Turkmenistan
11	20	120	–	–	–	25	–	–	Azerbaijan

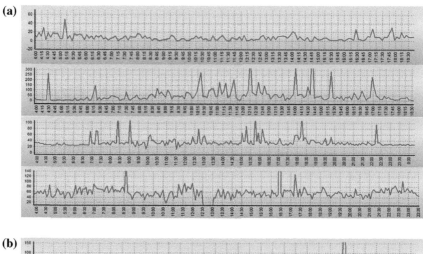

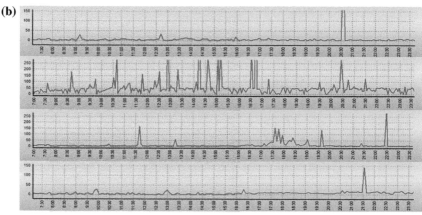

Fig. 8.8 **a** VII—Siazan, Qum Island, Shirvan, Neftchala: 2013-03-26 Georgia-Russia. **b** VII—Siazan, Naftalan, Shirvan, Qum Island 2013-05-27 2013-05-28 00:09:52.0 UTC mb 5.2 Georgia (Sak'art'velo). **c** VII—Qazakh, Siazan, Qum Island, Neftchala 2013-09-16 42.17 N; 45.89 E Russia. **d** X—Qum Island, Neftchala, Nakhchivan 2013-11-21 Iran-Iraq border. **e** III—Siazan, Neftchala, Qum Island, Turkmen01 2014-01-09 Caspian Sea, Offshore Azerbaijan. **f** I—Siazan, Neftchala, Qum Island, Turkmen01 2014-01-13 Turkmenistan. **g** IX—Qum Island, Shirvan, Nakhchivan, Neftchala 2014-01-28, 2014-01-28 23:47:38.0 UTC ML 5.1 WESTERN IRAN. **h** IV—Qum Island, Shirvan, Qazakh,Nakhchivan 2014-02-09, 2014-02-10 12:06:48.0 UTC Mw 5.4 AZERBAIJAN. **i** I—Siazan, Qum Island, Cybernetic, Neftchala, Shirvan, Turkmen01 2014-06-06, 2014-06-07 06:05:32.1 UTC mb 5.6 CASPIAN SEA, OFFSHR TURKMENISTAN. **j** Siazan, Qum Island, Shirvan, Qazakh 2014-06-29, 2014-06-29 17:26:10.4 UTC mb 5.1 AZERBAIJAN

8.3 Intelligent Seismic-Acoustic System for Identifying the ... 171

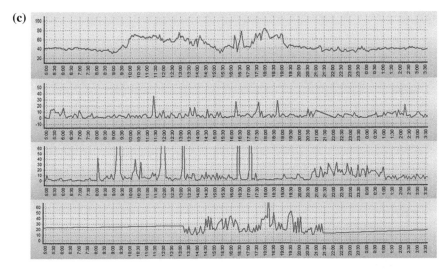

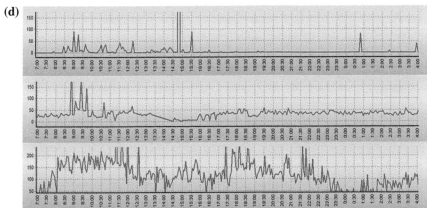

Fig. 8.8 (continued)

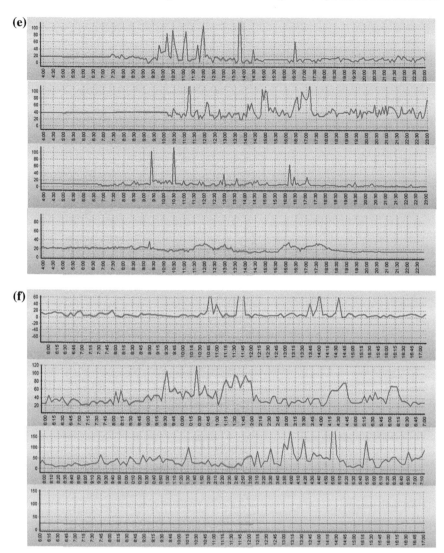

Fig. 8.8 (continued)

8.3 Intelligent Seismic-Acoustic System for Identifying the ...

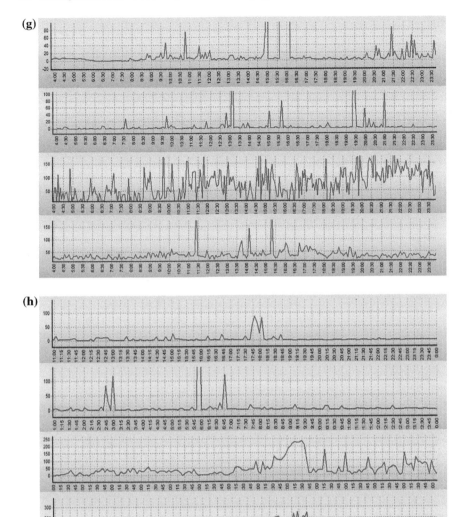

Fig. 8.8 (continued)

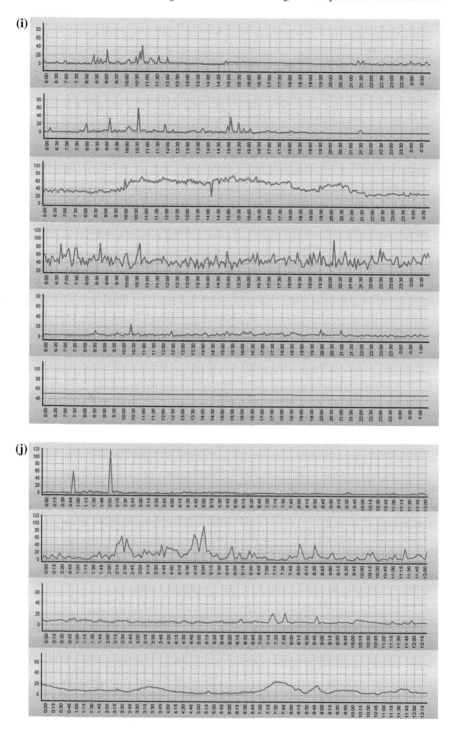

Fig. 8.8 (continued)

Row 1 of Table 8.2 shows the results of the identification of the area of the earthquake that occurred in Georgia on 2013.03.26. It follows from Fig. 8.8a that its beginning was indicated by the RNM ASP stations in the following order: Siazan—04:15; Qum Island—04:30; Shirvan—06:50; Neftchala—08:30. The system identified that such an indication corresponded to area VII. The indication was 8–10 h in advance of the earthquake.

Row 2 of Table 8.2 gives the results of the identification of the area of the earthquake that occurred in Georgia on 2013.05.27, 2013.05.28. According to the chart, the RNM ASP stations Siazan, Naftalan, Shirvan and Qum Island registered the ASP initiation more than 20 h before the earthquake. It is clear from the charts that the northern (Siazan) and north-western (Qum Island) stations detected the anomaly earlier than the rest of the stations. The beginning of ASP was indicated by the RNM ASP stations in the following order: Naftalan—07:30; Siazan—09:10; Shirvan—09:45; Qum Island—11:40 (Fig. 8.8b). Thus, the earthquake area was identified by the system by approximately 18:00 Baku time, which is almost 10–11 h earlier than the time of registration by the ground stations.

Rows 3 and 4 of Table 8.2 show the results of the identification of the area of the earthquakes that occurred in South Russia on 2013.09.16.

According to the charts of the third and fourth earthquakes (Fig. 8.8c), the ASP originated in the south-east of the Caucasus region and registered in the following order: Siazan—05:30, Qum Island—08:00, Qazakh—10:00, Neftchala—14:30. Based on this sequence of registration times, the system identified earthquake area VII, which corresponds to the north-east of Azerbaijan, where an earthquake actually occurred at 16:00–17:00 Baku time. The time of the area identification was approximately 15 h in advance of the earthquake.

According to the chart in Fig. 8.8d, based on the combination of times of registration by the stations Nakhchivan—08:00; Qum Island—09:00; Neftchala—08:50, the system identified the area of the expected earthquake on the Iran–Iraq border 12 h in advance.

Figure 8.8e shows the results of the identification of the area of the earthquake that occurred on 2014-01-09, at about 12:00, in the Caspian Sea, Offshore Azerbaijan, and was registered by the stations Turkmen01—09:15, Qum Island—09:25, Siazan—09:45, Neftchala—11:15, 16 h in advance of the earthquake.

Row 7 of Table 8.2 shows that the system identified the area of the expected earthquake in Turkmenistan. According to the chart in Fig. 8.8f, based on its sequence of registrations by the stations Neftchala—09:30, Siazan—10:45, Qum Island—11:30, the system demonstrated that the location of the expected earthquake was in Turkmenistan, i.e. in area I east of Azerbaijan. The area of the expected earthquake was identified more than 24 h before the earthquake was registered.

Row 8 of Table 8.2 shows the results of the identification of the area of the earthquake that occurred in Western Iran. According to Fig. 8.8g, the sequence of registrations by the stations Qum Island—09:45, Shirvan—07:30, Nakhchivan—04:50 and Neftchala—11:20 allowed the system to identify the location of the earthquake in area IX—Western Iran.

Row 9 of Table 8.2 shows the results of the identification of the area of the earthquake that occurred in Azerbaijan. According to the charts in Fig. 8.8h, the stations registered the corresponding ASP in the following order: Qum Island—17:45, Shirvan—12:45, Qazakh—19:00 and Nakhchivan—18:30, which allowed the system to identify the number (IV) of the area of the expected earthquake with a lead time of 19 h.

Row 10 of Table 8.2 shows the results of identification of the area of the earthquake that occurred in Offshore Turkmenistan. According to the charts in Fig. 8.8i, this event was registered by the stations Neftchala—06:45, Qum Island—07:55, Siazan—08:15, Cybernetic—09:55, Shirvan—10:20.

Row 11 of Table 8.2 contains the information on the identification of the location of the expected earthquake that occurred in area I in Azerbaijan. It follows from Fig. 8.8j that the ASP was indicated by the stations in the following order: Siazan—00:50, Qum Island—02:15, Shirvan—07:20, Qazakh—07:30, which allowed the system to identify the location of the area of the expected earthquake (Azerbaijan).

The analysis of the recorded data, some of which was presented above, has shown that in most cases ASP occurred in the areas where potential seismically active zones were identified based on the results of registration by ground seismic stations over a long period of time. With this in mind, it is obvious that when identifying the area of the focus of an expected earthquake by means of the readings of combinations of RNM ASP stations, it is necessary to take into account the information received by ground seismic stations in the past. An analysis has shown that the use of this additional information significantly increases the reliability of the results of the determination of the area of the focus of an expected earthquake.

After obtaining the above results by means of the existing RNM ASP stations, we built three more stations:

- in 2015, in the village of Lagich, Ismayilli region (seismically active zone—the southern border of the Republic of Dagestan, Russian Federation);
- in 2016 in the Lerik region (the northern border with the Islamic Republic of Iran);
- in 2017 in Sheki region (seismically active zone bordering with the Mingachevir reservoir).

All these stations are built on 40–100 m deep wells. Our experiments after the inclusion of additional seismic-acoustic stations in the network showed the possibility of using the intelligent seismic-acoustic system in the warning mode to inform about the area of the focus of the expected earthquake. By disconnecting the power and gas supply systems to ensure fire safety and evacuating population from earthquake-prone buildings and structures, it is possible to significantly reduce the damage and losses from possible earthquakes.

On the basis of the experimental studies on the development and use of networks of seismic-acoustic stations for the registration of seismic processes prior to earthquakes conducted from 2010 to 2017, the following can be pointed out:

1. The intelligent system consisting of the network of the RNM ASP stations and an expert system combined with a neural network can be used as a toolkit for

identifying the location of the area of an expected earthquake in advance. The seismologist is supplied with the information containing the direction and the number of the area of the focus of an expected earthquake, current combinations of the parameters of ASP, and the amount, list, date and magnitude of similar combinations registered in the given area during previous earthquakes, as well as the information on the seismic activity based on the readings of ground seismic stations in the area. This information will allow the seismologist, after he/she gains certain amount of experience, to evaluate the degree of authenticity of the obtained information on the location of the area of the focus of an expected earthquake. Having enough time before the earthquake occurs, the seismologist can involve other specialists in the decision-making process if there is any doubt, ruling out possible errors.

2. The RNM ASP stations in the proposed network are built on wells with different depths and, consequently, different characteristics. These characteristics are difficult to take into account in identifying the location of the area of the focus of an expected earthquake and in determining its magnitude.

    Moreover, as the depth of a well increases, its cost increases sharply, too. For instance, the drilling a 4000–5000 m deep well can cost up to 30 million dollars. This makes the building of the RNM ASP stations in the countries that have no suspended oil wells quite challenging. In this case, we recommend forming a network of stations built in 50–100-m deep water wells in the future, with hydrophones placed in the water column at a depth of 10–20 m. To improve the validity and reliability of the identification of the location of the area of the expected earthquake, we found it appropriate to build a network consisting of a large number of stations (over 10–15) in wells of equal depth located at an equal distance from one another. An integration of networks of RNM ASP stations of the countries in several seismically active regions via satellite communication can, in the long term, allow increasing the validity and reliability of determining the coordinates of the location of the focus of an expected earthquake.

3. Our experiments have demonstrated that the reliability of the ASP monitoring results and the validity of the results of identification of the location of the area of the focus of an expected earthquake grow with the growth in earthquake strength. With the strength exceeding 5 points, the results of the identification of the earthquake location proved to be valid in almost all cases. The value of the estimate of the cross-correlation function $R_{X\varepsilon}(\mu)$ between the useful signal $X(i\Delta t)$ and the noise $\varepsilon(i\Delta t)$ decreases as the distance from the earthquake area grows. In contrast, the value of the estimate of noise variance D_ε, increases as the distance from the area grows. The propagation velocity of the seismic-acoustic noise in different types of medium, e.g. water, sand or clay, significantly varies. There is a correlation between the well depth and the radius of ASP monitoring.

4. The experiments at the Qum Island station in the Caspian Sea have demonstrated that the monitoring range of that station is much wider than that of the stations located far from the Caspian Sea. Other stations in Siazan and Neftchala located near the Caspian Sea also have a wide monitoring range compared with other stations. Practically all seismic processes reaching the Caspian Sea are clearly

registered by those stations. Therefore, in building networks of new stations, one should consider the fact that the sea is a "perfect conductor" for seismic-acoustic noises emerging during ASP initiation in the region.

5. The results obtained from the experimental data give us reason to assume that the lead time of the registration of the ASP initiation by a seismic-acoustic RNM ASP station over standard seismic equipment is due to two factors.

   First, seismic-acoustic waves that arise in the beginning of the ASP initiation do not reach the earth's surface due to the frequency characteristics of certain upper strata, which furthers their horizontal spread in deep strata as noise. Reaching the steel pipes of the well, seismic-acoustic waves transform into acoustic signals and go to the ground surface at the velocity of sound, where they are detected by a hydrophone. At the same time, low frequency seismic waves from seismic processes are perceived at the surface after a certain amount of time, when the earthquake is already in progress, and are registered by seismic receivers of standard ground equipment much later.

   Second, the use of noise technologies by analyzing seismic-acoustic noise allows, when a correlation appears between the useful signal and the noise, registering ASP at the beginning of their initiation.

   These two factors make it possible for RNM ASP stations to indicate the time of the beginning of the ASP initiation much earlier than is done by stations of the seismological service.

6. Seismic-acoustic stations of ASP monitoring can also be used for monitoring the latent period of volcano formation well before eruption. Their use will also allow performing the monitoring, on a regional scale, of the tests of minor and major nuclear bombs and other experiments related to the manufacture of military equipment and weapons.

References

1. Kanamori H, Brodsky EE (2004) The physics of earthquakes. Rep Prog Phys 67(8):1429–1496. https://doi.org/10.1088/0034-4885/67/8/R03
2. Tothong P, Cornell CA (2006) An empirical ground-motion attenuation relation for inelastic spectral displacement. Bull Seismol Soc Am 96(6):2146–2164. https://doi.org/10.1785/0120060018
3. Ghahari SF, Jahankhah H, Ghannad MA (2010) Study on elastic response of structures to near-fault ground motions through record decomposition. Soil Dyn Earthq Eng 30(7):536–546. https://doi.org/10.1016/j.soildyn.2010.01.009
4. Boore DM, Bommer JJ (2005) Processing of strong-motion accelerograms: needs, options and consequences. Soil Dyn Earthq Eng 25(2):93–115. https://doi.org/10.1016/j.soildyn.2004.10.007
5. Galiana-Merino JJ, Parolai S, Rosa-Herranz J (2011) Seismic wave characterization using complex trace analysis in the stationary wavelet packet domain. Soil Dyn Earthq Eng 31(11):1565–1578. https://doi.org/10.1016/j.soildyn.2011.06.009
6. Yee E, Stewart JP, Schoenberg FP (2011) Characterization and utilization of noisy displacement signals from simple shear device using linear and kernel regression methods. Soil Dyn Earthq Eng 31(1):25–32. https://doi.org/10.1016/j.soildyn.2010.07.011

References

7. Pavlović VD, Veličković ZS (1998) Measurement of the seismic waves propagation velocity in the real medium. Sci J Facta Univ Ser: Phys Chem Technol 1:63–73
8. Wang Y, Lu J, Shi Y et al (2009) PS-wave Q estimation based on the P-wave Q values. J Geophys Eng 6(4):386–389. https://doi.org/10.1088/1742-2132/6/4/006
9. Mallat SG (1989) A theory for multiresolution signal decomposition: the wavelet representation. IEEE Trans Pattern Anal Mach Intell 11(7):674–693. https://doi.org/10.1109/34.192463
10. Vidakovic B, Lozoya CB (1998) On time-dependent wavelet denoising. IEEE Trans Sig Process 46(9):2549–2554. https://doi.org/10.1109/78.709544
11. Colak OH, Destici TC, Ozen S et al (2008) Frequency-energy characteristics of local earthquakes using discrete wavelet transform (DWT). Int J Civil Environ Eng 2(8):199–202
12. Hutton DV (2004) Fundamentals of finite element analysis, 1st edn. McGraw-Hill, New York
13. Kislov KV, Gravirov VV (2011) Earthquake arrival identification in a record with technogenic noise. Seismic Instrum 47(1):66–79. https://doi.org/10.3103/S0747923911010129
14. Descherevsky AV, Lukk AA, Sidorin AY et al (2003) Flicker-noise spectroscopy in earthquake prediction research. Nat Hazards Earth Syst Sci 3(3–4):159–164. https://doi.org/10.5194/nhess-3-159-2003
15. Kossobokov VG (2006) Testing earthquake prediction methods: «The West Pacific short-term forecast of earthquakes with magnitude MwHRV ≥ 5.8». Tectonophysics 413(1–2):25–31. https://doi.org/10.1016/j.tecto.2005.10.006
16. Shebalin P, Keilis-Borok V, Gabrielov A et al (2006) Short-term earthquake prediction by reverse analysis of lithosphere dynamics. Tectonophysics 413(1–2):63–75. https://doi.org/10.1016/j.tecto.2005.10.033
17. Hashemi M, Alesheikh AA (2011) A GIS-based earthquake damage assessment and settlement methodology. Soil Dyn Earthq Eng 31(11):1607–1617. https://doi.org/10.1016/j.soildyn.2011.07.003
18. Hatzigeorgiou GD, Beskos DE (2010) Soil–structure interaction effects on seismic inelastic analysis of 3-D tunnels. Soil Dyn Earthq Eng 30(9):851–861. https://doi.org/10.1016/j.soildyn.2010.03.010
19. Papagiannopoulos GA, Beskos DE (2006) On a modal damping identification model for building structures. Arch Appl Mech 76:443–463
20. Papagiannopoulos GA, Beskos DE (2009) On a modal damping identification model for non-classically damped linear building structures subjected to earthquakes. Soil Dyn Earthq Eng 29(3):583–589
21. Zafarani H, Noorzad A, Ansari A et al (2009) Stochastic modeling of Iranian earthquakes and estimation of ground motion for future earthquakes in Greater Tehran. Soil Dyn Earthq Eng 29(4):722–741. https://doi.org/10.1016/j.soildyn.2008.08.002
22. Sokolov VY, Loh CH, Wen KL (2003) Evaluation of hard rock spectral models for the Taiwan region on the basis of the 1999 Chi-Chi earthquake data. Soil Dyn Earthq Eng 23(8):715–735. https://doi.org/10.1016/S0267-7261(03)00075-7
23. Stankiewicz J, Bindi D, Oth A et al (2013) Designing efficient earthquake early warning systems: case study of Almaty, Kazakhstan. J Seismolog 17(4):1125–1137. https://doi.org/10.1007/s10950-013-9381-4
24. Rydelek P, Pujol J (2004) Real-time seismic warning with a two-station subarray. Bull Seismol Soc Am 94(4):1546–1550. https://doi.org/10.1785/012003197
25. Aliev TA, Abbasov AM, Aliev ER et al (2007) Digital technology and systems for generating and analyzing information from deep strata of the earth for the purpose of interference monitoring of the technical state of major structures. Autom Control Comput Sci 41(2):59–67. https://doi.org/10.3103/S0146411607020010
26. Aliev TA, Alizada TA, Abbasov AM (2009) Method for monitoring the beginning of anomalous seismic process. Eurasian Patent 012,803, 30 Dec 2009
27. Aliev T, Quluyev Q, Pashayev F et al (2016) Intelligent seismic-acoustic system for identifying the location of the areas of an expected earthquake. J Geosci Environ Prot 4(4):147–162. https://doi.org/10.4236/gep.2016.44018

28. Aliev TA, Abbasov AM, Guluyev QA et al (2013) System of robust noise monitoring of anomalous seismic processes. Soil Dyn Earthq Eng 53:11–25. https://doi.org/10.1016/j.soildyn.2012.12.013
29. Aliev TA (2017) Intelligent seismic-acoustic system for identifying the area of the focus of an expected earthquake. In: Taher Zouaghi (ed) Earthquakes: tectonics, hazard and risk mitigation. Intech, London. pp 293–315. https://doi.org/10.5772/65403
30. Aliev TA, Aliev ER (2008) Multichannel telemetric system for seismo-acoustic signal interference monitoring of earthquakes. Autom Control Comput Sci 42(4):223–228. https://doi.org/10.3103/S0146411608040093
31. Aliev TA, Musayeva NF, Sattarova UE (2014) Noise technologies for operating the system for monitoring of the beginning of violation of seismic stability of construction objects. In: Zadeh L., Abbasov A., Yager R. et al (eds) Recent developments and new directions in soft computing. Studies in fuzziness and soft computing, vol 317. Springer, Cham, pp 211–232. https://doi.org/10.1007/978-3-319-06323-2_14
32. Aliev TA, Alizade AA, Etirmishli GD et al (2011) Intelligent seismoacoustic system for monitoring the beginning of anomalous seismic process. Seismic Instrum 47(1):15–23. https://doi.org/10.3103/S0747923911010026
33. Aliev TA, Abbasov AM, Mamedova GG et al (2013) Technologies for noise monitoring of abnormal seismic processes. Seismic Instrum 49(1):64–80. https://doi.org/10.3103/S0747923913010015
34. Aliev TA (2000) Robust technology for systems analysis of seismic signals. Autom Control Comput Sci 34(5):17–26
35. Aliev TA, Abbasov AM (2005) Digital technology and a system of interference monitoring of the technical state of physical structures and warnings of anomalous seismic processes. Autom Control Comput Sci 39(6):1–7

Chapter 9
Possibilities of Using Noise Control Technology in Medicine

Abstract The possibility of using laptops and smartphones to control the state of the heart by analyzing the noise of its sounds is considered. The results of heart control experiments using the proposed noise technologies are presented. The experiments were carried out by means of a Lithmann 3200 stethoscope connected to a laptop via Bluetooth. It is shown that the results of numerous experiments on monitoring the change in the state of heart activity, some of which are given in the corresponding tables, confirmed the expediency of using the proposed technology by users of laptops and smartphones. To enhance the reliability of these results, technology has also been developed to monitor the state of the heart based on the spectral characteristics of the noise of its sounds. Experiments have shown that by duplicating these technologies it is possible to enhance the reliability and validity of monitoring results.

9.1 Possibilities of Using Laptops and Smartphones for the Control of the State of the Heart

More and more people from various strata of society suffer from heart diseases these days. Modern diagnostics of these diseases is further complicated by physicians' heavy workload and the high cost of most medical diagnostic methods. In many cases, people suffering from cardiovascular diseases only visit a cardiologist when the disease is already in its explicit, expressed form, and even though the disease is diagnosable by the known methods, its treatment becomes substantially complicated [1, 2].

It is a fact that an untimely diagnosed heart disease can cause tragic consequences for many patients, which is why diagnostics of these diseases are in the focus of attention of many researchers and physicians [2–17].

Existing mobile means of monitoring do not permit online assessment of the condition of the cardiovascular system (CVS). Some systems solve the task at hand by transmitting ECG to the server via communication channels for subsequent processing and analysis at a cardiology center [2–17]. However, the procedure in question is quite expensive, which impedes the possibility of its mass use.

The well-known Holter monitor continuously records ECG in the course of 24 h for its subsequent analysis by a cardiologist [3]. This system detects changes in the condition of CVS only after a certain period of time and requires involvement of a cardiologist, which can delay monitoring results.

Another system, "Easy ECG Mobile", records ECG and the record is instantly transmitted via a wireless local area network or within several seconds via the Internet [4].

"Ericsson Mobile Health" (EMH) system allows for remote monitoring of the condition of CVS.

There are number of works [11, 17–24] dealing with this problem.

The described mobile monitoring tools have the following shortcomings:

The described mobile tools for monitoring the state of the cardiovascular system have the following shortcomings:

- inability to assess the need for a medical examination online and without the involvement of medical personnel;
- inability to carry out continuous monitoring of changes in the condition of cardiovascular patients;
- inability to detect the initial stage of cardiovascular diseases without the involvement of medical personnel, which might help people avoid further complications if they contacted a physician in time.

In view of the above, the issue of interest here is online identification of the beginning of changes in the heart activity based on the characteristics of heart sounds, using laptops or smartphones. This will allow potentially sick people to make sure in home conditions that they should see a doctor. Otherwise, they will be informed based on the monitoring results that there is no need to do so, thereby minimizing the number of unwarranted visits to a doctor and saving medical professionals' time. Patients, in turn, will be able to avoid unnecessary waste of time, energy and finances. None of the systems described in the previous chapters can perform this function online and without medical personnel. This chapter is devoted to one method of solving solve this problem.

In this chapter, we propose technologies for monitoring the beginning of changes in CVS by identifying heart sounds with the use of laptops and smartphones. This will allow users of laptops and smartphones to receive information on the condition of their heart in real time, in home or work conditions.

This will allow a user to determine if a visit to a doctor is necessary, using only a laptop or a smartphone.

As was noted earlier, timely diagnostics of cardiovascular diseases today is impeded by high cost of medical examinations. Besides, the procedure takes a lot of patients' and doctors' time and energy.

With this system, using a laptop or a smartphone, one can ascertain in home conditions that no change has occurred in one's heart activity and that an expensive, lengthy and exhausting preventive medical examination is unnecessary. In other cases, the proposed system will allow for monitoring of the initial stage, when the disease is easily treatable [1, 23, 24].

9.1 Possibilities of Using Laptops and Smartphones …

In view of the above, the development of such systems is of very important practical interest [17–26]. This chapter deals with the possibility of the development and the prospects of mass use of mobile tools for monitoring of such widespread diseases as cardiovascular diseases. For this purpose, we can use heart sounds to evaluate a person's health in the monitoring of changes in the heart activity. However, the error in the results of the analysis of the effects of the noise $\varepsilon(t)$ of the heart sounds varies across a very wide range. This is because auscultation can be performed in home or industrial conditions, at considerable temperature fluctuations depending on the season, in windy or rainy weather, vibration, shaking, extraneous noises, etc., are also possible.

As a result, during auscultation, a signal contaminated with the noise $\varepsilon(t)$ arrives at the microphone input instead of the useful signal $X(t)$. The analyzed signal looks as follows in the analog form:

$$g(t) = X(t) + \varepsilon(t),$$

and as follows in the digital form:

$$g(i\Delta t) = X(i\Delta t) + \varepsilon(i\Delta t).$$

Due to the abovementioned reasons, both the amplitude and the spectrum of the noise $\varepsilon(i\Delta t)$ vary across a rather wide range. For the same reasons, the errors of the obtained estimates of the correlation functions $R_{gg}(i\Delta t)$ of the heart sound $g(i\Delta t)$ also vary across a wide range. Thus, we fail to provide the condition of robustness for the estimates of the correlation function in real-time mode, i.e., to rule out the dependence of the obtained results on the variation of the characteristics of the noise $\varepsilon(i\Delta t)$. This, in turn, complicates solving the problem of identification of heart sounds by correlation methods. Consequently, ensuring an adequate identification requires the elimination of the effects of the said factors on the errors in the estimates $R_{gg}(i\Delta t)$ [1, 27–30].

At first glance, the effects of the said errors on the results of identification of heart sounds can be eliminated by filtering the noise accompanying the useful signal $X(i\Delta t)$. If the noise spectrum is stable, filtration usually gives satisfactory results. Under the field conditions, however, the spectrum of the noise and its variance varies across a very wide range, and we cannot get the desired accuracy of calculations using filtration technology. Thus, we cannot always achieve satisfactory results through correlation analysis of heart sounds using filtration. Therefore, solving the problem under consideration first requires developing the technologies that can calculate such estimates of correlation characteristics that remain practically unaffected by changes in the said noises [1, 2, 27–30].

To that end, it is appropriate to first reduce the estimates $R_{gg}(i\Delta t)$ to a single dimensionless value by applying the normalization procedure [1, 2, 28, 29]. Our analysis, however, demonstrates that the application of traditional methods introduces an additional error into the normalized estimates of the correlation functions

$r_{gg}(i\Delta t)$, which, in turn, also complicates attempts to ensure the adequacy of the results of heart sounds analysis.

As was indicated in Chap. 3, a normalization of correlation functions gives the correct result only at $\mu = 0$. For all other cases, when $\mu \neq 0$, the results of normalization of the correlation functions of the noisy signal $g(i\Delta t)$ differ from the results of normalization of the correlation functions of the useful signal $X(i\Delta t)$. Our experiments demonstrate that this is the main factor preventing the adequacy of the results of identification of heart sounds. Therefore, to eliminate the errors of the traditional normalization of the estimates of $r_{gg}(\mu \neq 0)$, this error should be corrected. Thus, to design a system for CVS monitoring, we have to develop a technology aimed at eliminating the effects of noise $\varepsilon(i\Delta t)$ on the estimates of the normalized correlation functions at $r_{gg}(\mu \neq 0)$ of the noisy signals $g(i\Delta t)$ so that the following condition is fulfilled:

$$r_{gg}(\mu) \approx r_{XX}(\mu).$$

To this end, the estimates of the normalized correlation function of the noisy signal $g(i\Delta t)$ at $\mu \neq 0$ should be calculated from the expression

$$r_{gg}(\mu \neq 0) = \frac{R_{gg}(\mu \neq 0)}{D_X}.$$

As the variance D_g of the noisy signal is the following sum [1, 2, 29, 30]

$$D_g = D_X + D_\varepsilon,$$

the formula for calculating the normalized correlation function can be represented as follows:

$$r_{gg}(\mu \neq 0) = \frac{R_{gg}(\mu \neq 0)}{(D_g - D_\varepsilon)}.$$

To implement this expression, we clearly need to calculate the estimate of the noise variance D_ε, the formula for which will be given in later paragraphs.

Thus, we can ensure the accuracy of the estimates of the normalized correlation functions by eliminating the additional noise-induced errors.

Proposed below is the appropriate sequence of calculations, the totality of which makes up the procedure of normalization of the correlation functions of the heart sound $g(i\Delta t)$.

(1) The estimates of the correlation function of the signal $g(i\Delta t)$ at $\mu = 1, 2, 3, 4, \ldots$ are calculated from the following expression:

$$R_{gg}(\mu) = \frac{1}{N}\sum_{i=1}^{N} g(i\Delta t)g((i+\mu)\Delta t).$$

(2) The estimate of the variance of the heart sound is calculated from the expression

$$D_g = \frac{1}{N} \sum_{i=1}^{N} g(i\Delta t) g(i\Delta t).$$

(3) The estimate of the variance D_ε of the noise of the heart sound is calculated from the expression

$$D_\varepsilon \approx \frac{1}{N} \sum_{i=1}^{N} \left[g^2(i\Delta t) + g(i\Delta t) g((i+2)\Delta t) - 2g(i\Delta t) g((i+1)\Delta t) \right].$$

(4) The value of the variance $R_{XX}(\mu = 0)$ of the useful signal $X(i\Delta t)$ is calculated from the expression

$$R_{UU}(\mu = 0) = D_g - D_\varepsilon.$$

(5) The estimates of the normalized correlation function $r_{gg}(\mu)$ of the heart sound $g(i\Delta t)$ at $\mu = 0, 1, 2, 3, \ldots$ are calculated from the expression

$$r_{gg}(\mu) = \begin{cases} \frac{R_{gg}(\mu=0)}{D_g=1} & \text{at } \mu \neq 0 \\ \frac{R_{gg}(\mu)}{(D_g - D_\varepsilon)} & \text{at } \mu \neq 0 \end{cases}$$

Our theoretical and experimental studies have demonstrated that when the proposed procedure is applied to calculating the corrected estimates of $r_{gg}(\mu)$ of the heart sound $g(i\Delta t)$, the shape of the curve of its normalized correlation function changes in relation to the changes in the state of heart activity. This specific characteristic of the corrected normalized correlation functions of the heart sound enabled us to create a technology for identifying the state of the heart [1, 2, 29, 30].

Our experiments have demonstrated that the estimates $R^*_{X\varepsilon}(1\Delta t)$, $R^*_{X\varepsilon}(2\Delta t)$, $R^*_{X\varepsilon}(3\Delta t)$, ..., which change as the state of the heart changes, should be also used as informative attributes [1, 2].

This is due to that fact that the sum noise, i.e., the estimate of D_ε, is also affected by the value of the cross-correlation function between the useful signal $g(i\Delta t)$ and the noise $\varepsilon(i\Delta t)$. According to our research, certain information about it can be obtained from the estimate of the relay cross-correlation function between $X(i\Delta t)$ and $\varepsilon(i\Delta t)$, which can be calculated from the expression

$$R^*_{U\varepsilon} = \frac{1}{N} \sum_{i=1}^{N} [\operatorname{sgn} g(i\Delta t) g(i\Delta t) - 2\operatorname{sgn} g(i\Delta t) g((i+1)\Delta t) + \operatorname{sgn} g(i\Delta t) g((i+2)\Delta t)].$$

Thus, when this technology is used, every time after an auscultation of the heart by a microphone, combinations of values of $r_{gg}(\mu)$ at $\mu_{0.75}$, $\mu_{0.5}$, $\mu_{0.25}$, μ_0 are calculated as informative attributes, following the calculation of the estimates of the corrected normalized correlation function of the heart sound. It has been established through long-term experiments that these values practically contain the information on the dynamic of changes in the state of the heart because the difference of their combinations change with the changes in the state of the heart. This peculiarity of changes in the estimates of $r(\mu)$ during changes in the state of the heart allowed us to use them as indicators of the start of those changes. It is therefore obvious that when identifying the state of the heart activity, reference estimates for the corresponding states of the heart can be formed from combinations of the informative attributes $r(\mu)$ at $\mu_{0.75}$, $\mu_{0.5}$, $\mu_{0.25}$, μ_0 and $R_{X\varepsilon}(\mu = 1\Delta t)$ for every user. Comparing them with the current estimates, the monitoring of the state of the heart can be carried out in real time.

Numerous tests have been conducted to validate the reliability and accuracy of the results obtained by means of this technology. The program of the procedure of monitoring of changes in the state of the heart by analyzing its sound $g(i\Delta t)$ has been created in Matlab.

The experiments were conducted over 20 days on three groups of researchers. The first group included young people of 20–30 years old. The second group was comprised of middle-aged people—30–50 year old, and the third one of seniors, aged 70–80.

Fifteen persons from these age groups were tested daily for 20 days. Auscultation of each participant's heart sounds was performed in two stages: first at rest, then after a certain physical stress. The estimates were calculated before the physical stress and immediately after. It is well known that physical stress causes changes in the state of the heart compared with its state at rest. We can, therefore, assume that if current combinations of informative attributes differ from the reference ones, the proposed technology can be supposedly used to identify any change in the state of the heart. For the experiments to be more informative, the participants of the test were exposed to various kinds of exercise, such as sit-ups, jogging. Combinations of informative attributes in the form of $R_{X\varepsilon}(\mu)$ and $r(\mu)$ at $\mu_{0.75}$, $\mu_{0.5}$, $\mu_{0.25}$, μ_0 were formed based on the obtained results. Changes in the current estimates of these informative attributes after a variety of physical exercises have allowed us to assess the validity and reliability of monitoring of changes in the state of the heart using heart sounds. Tables 9.1, 9.2, 9.3, 9.4, 9.5, 9.6 contain the results of the experiment for different age groups on one day only, as the results of the other days are just insignificantly different. It is clear from these tables that using combinations of $R_{X\varepsilon}(\Delta t)$ and $r_{gg}(\mu)$ at $\mu_{0.75}$, $\mu_{0.5}$, $\mu_{0.25}$, μ_0, we can perform identification of the state of anyone's heart activity in real time. Numerous experiments, some of which are presented in Tables 9.1, 9.2, 9.3, 9.4, 9.5, 9.6, confirmed this specific characteristic of these informative attributes. As these informative attributes are easy to calculate, they can be easily implemented by means of laptops and smartphones, which opens the way for wide universal application of the technology in question by users of laptops and smartphones.

It should be noted that in our experiments, we sampled the heart sound $g(i\Delta t)$ at the interval $\Delta t = 1$ ms. During the calculation of the estimates of $r_{gg}(\mu)$, the calculations were conducted first at

9.1 Possibilities of Using Laptops and Smartphones …

Table 9.1 Monitoring results for Turana Aliyeva, 22 y/o

Date and time	State	$\mu_{0.75}$	$\mu_{0.5}$	$\mu_{0.25}$	μ_0	$R_{X\varepsilon}(\Delta t)$
04.05.15 10:37	A.R.[a]	25	35	45	50	−0.0377
	A.S.[b]	20	30	35	35	−0.0559

[a] At rest
[b] After stress

Table 9.2 Monitoring results for Narmin Rzayeva, 25 y/o

Date and time	State	$\mu_{0.75}$	$\mu_{0.5}$	$\mu_{0.25}$	μ_0	$R_{X\varepsilon}(\Delta t)$
05.05.15 12:03	A.R.	25	35	40	45	−0.0393
	A.S.	20	25	30	30	−0.0590

Table 9.3 Monitoring results for Zarina Bashirova, 30 y/o

Date and time	State	$\mu_{0.75}$	$\mu_{0.5}$	$\mu_{0.25}$	μ_0	$R_{X\varepsilon}(\Delta t)$
04.05.15 10:38	A.R.	25	35	45	60	−0.0532
	A.S.	20	30	35	35	−0.0718

Table 9.4 Monitoring results for Natalya Belyayeva, 40 y/o

Date and time	State	$\mu_{0.75}$	$\mu_{0.5}$	$\mu_{0.25}$	μ_0	$R_{X\varepsilon}(\Delta t)$
04.05.15 11:21	A.R.	25	35	40	50	−0.0458
	A.S.	20	30	35	40	−0.0602

Table 9.5 Monitoring results for Sadagat Gurbanova, 58 y/o

Date and time	State	$\mu_{0.75}$	$\mu_{0.5}$	$\mu_{0.25}$	μ_0	$R_{X\varepsilon}(\Delta t)$
04.05.15 12:17	A.R.	30	40	55	70	−0.0281
	A.S.	30	40	50	60	−0.0414

Table 9.6 Monitoring results for Telman Aliev, 80 y/o

Date and time	State	$\mu_{0.75}$	$\mu_{0.5}$	$\mu_{0.25}$	μ_0	$R_{X\varepsilon}(\Delta t)$
05.05.15 11:49	A.R.	25	35	50	70	−0.0266
	A.S.	25	35	45	55	−0.0361

$$\mu_1 = 1\Delta t, \ \mu_2 = 2\Delta t, \ \mu_3 = 3\Delta t, \ldots,$$

then at

$$\mu_1 = 2\Delta t, \ \mu_2 = 4\Delta t, \ \mu_3 = 6\Delta t, \ldots,$$

then at

$$\mu_1 = 3\Delta t, \ \mu_2 = 6\Delta t, \ \mu_3 = 9\Delta t, \ \ldots,$$

then at

$$\mu_1 = 4\Delta t, \ \mu_2 = 8\Delta t, \ \mu_3 = 12\Delta t, \ \ldots,$$

and finally at

$$\mu_1 = 5\Delta t, \ \mu_2 = 10\Delta t, \ \mu_3 = 15\Delta t, \ \ldots.$$

Given below are the results for the case when

$$\mu_1 = 5\Delta t, \ \mu_2 = 10\Delta t, \ \mu_3 = 15\Delta t, \ \ldots.$$

Our experiments have established that the shape of the normalized correlation function for a healthy person in the normal state of heart activity is an exponential curve. This shape changes at the start of physical stress and approaches gradually to a linear form as the stress grows. It is almost linear at heavy stress. Comparing Tables 9.1–9.3 and Tables 9.4–9.5, we can see that the transition from an exponential form to a linear form is much faster for young people than for older people. It was obvious that by identifying the heart sound from the combinations of the informative attributes at $\mu_{0.75}$, $\mu_{0.5}$, $\mu_{0.25}$, μ_0 obtained from the estimates of the normalized correlation functions before and after a 2–3 min long physical exercise, we could identify even an insignificant change in the state of the heart. Therefore, it is clearly possible to perform reliable monitoring of changes in the state of CVS in real time in home or work conditions.

In home or work conditions this can be implemented in several ways. The first option is to perform monitoring by means of Lithmann 3200 electronic stethoscope connected to a laptop via Bluetooth (Fig. 9.1). T high sensitivity of this stethoscope allows one to easily listen to heart sounds. The second option is to perform monitoring by means of a smartphone when its microphone is used to listen to heart sounds (Fig. 9.1). Both ways to perform the procedure of auscultation and monitoring of the state of the heart are given in Fig. 9.1.

Naturally, in both versions, the problem of identification of the state of the heart is solved by means of software tools.

The first stage of the process of training to monitor the condition of CVS with the use of a laptop or a smartphone the system is as follows. Every day at home or at work, in the most indicative periods of the day, the owner of the laptop/smartphone places the stethoscope diaphragm/smartphone microphone on a certain spot in the cardiac area, as Fig. 9.1 shows. It should be noted that samples of the heart sound $g(i\Delta t)$ are measured at the interval $\Delta t = 1$ ms, i.e., the signal $g(i\Delta t)$ obtained at the output of the microphone is sampled with the frequency 1 kHz. Thus, the digital

9.1 Possibilities of Using Laptops and Smartphones …

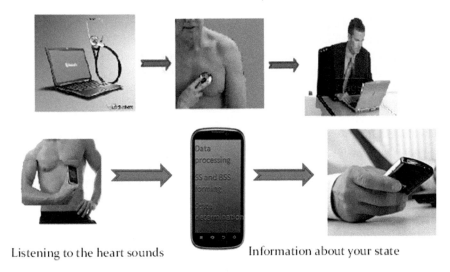

Fig. 9.1 Concept of two versions of the system for online monitoring of changes in CVS

information from the stethoscope reflecting the state of CVS is saved into a laptop or smartphone memory as a file. Then the estimates of the normalized correlation functions of the heart sound are calculated and the combinations of the informative attributes $\mu_{0.75}$, $\mu_{0.5}$, $\mu_{0.25}$, μ_0, $R_{X\varepsilon}$ that correspond to the current state of the heart activity are calculated by means of the technology described earlier [1, 2].

Thus, in the system training period, the owner may perform auscultation of his/her heart several times a day, starting in the morning. The aforementioned estimates form the combination of informative attributes corresponding to the user's current condition and they are included in the set of the morning condition as reference estimates. Similarly, reference estimates for other parts of the day are created. Our experiments have demonstrated that the estimates obtained in near time intervals will differ from one another only insignificantly. It is, therefore, appropriate to listen to heart sounds at intervals of at least one hour. As a result of the training, several sets will be formed for the owner during the day. Those sets form the bank of reference estimates, which are individual for this particular owner. The procedure is repeated in the following days, and if any combination of estimates matches the one already existing in the bank, the fact of the match is registered. If the current combination differs from all those present in the bank, it is entered into the corresponding set as a new reference estimate. Eventually, we will start obtaining current estimates that will be equal or close to the combinations from the corresponding set, which are already present in the bank. At that point, the training process will be considered complete [1, 2, 27–30].

From then on, the owner of the device continues performing auscultation at corresponding periods of the day in his/her everyday life. If a resulting combination of estimates matches the estimates from the corresponding set in the bank, the screen displays the following message: "No changes in the state of your heart". If the current combination of estimates in some period of time differs from all those corresponding to this state and already present in the bank, the device screen displays "There have been changes in the state of your heart" message and the list of all combinations of estimates relevant to this state from the bank, as well as the values of the estimates of the current combination. If the difference between those combinations is significant, the owner receives a recommendation to consult a cardiologist. If the cardiologist concludes after an examination that there is no reason to worry, that current combination is also included in the corresponding set of reference combinations and saved in the bank of reference estimates [1, 2].

The tests and experiments on the proposed technologies have been conducted, using Lithmann 3200 stethoscope connected to a laptop via Bluetooth. The results of the monitoring of changes in the state of heart activity, some of which are presented in Tables 9.1, 9.2, 9.3, 9.4, 9.5, 9.6, have demonstrated that it is possible to ensure the reliable and effective use of the system in question by laptop and smartphone users.

9.2 Technology for Monitoring the State of the Heart with the Use of the Spectral Characteristics of Heart Sounds

It is known that in the normal state of functioning of the heart, due to the lack of correlation between the useful signal $X(t)$ and the noise $\varepsilon(t) = \varepsilon_1(t)$, the following conditions are fulfilled [1–3]:

$$M[X(t)X(t)] \neq 0, \quad M[\varepsilon_1(t)X(t)] = 0,$$
$$M[X(t)\varepsilon_1(t)] = 0, \quad M[\varepsilon_1(t)\varepsilon_1(t)] \neq 0,$$
$$D_g = M[X(t)X(t) + \varepsilon_1(t)\varepsilon_1(t)].$$

As was discussed earlier, as the functioning of the heart changes, in addition to the noise $\varepsilon_1(i\Delta t)$, the noise $\varepsilon_2(i\Delta t)$ correlated with the useful signal $X(t)$ [1, 29] emerges in the heart sound $g(t)$. $\varepsilon_2(i\Delta t)$ contains the information about the changes in the state of the heart [1, 2]. Thus, due to the correlation between the useful signal $X(t)$ and the sum noise $\varepsilon(i\Delta t) = \varepsilon_1(i\Delta t) + \varepsilon_2(i\Delta t)$, the indicated conditions are violated, because the following takes place

$$D_g = M[(X(t) + \varepsilon_1(t) + \varepsilon_2(t))(X(t) + \varepsilon_1(t) + \varepsilon_2(t))]$$
$$= M[X(t)X(t)] + M[X(t)\varepsilon_1(t)] + M[X(t)\varepsilon_2(t)] + M[\varepsilon_1(t)X(t)]$$
$$+ M[\varepsilon_1(t)\varepsilon_1(t)] + M[\varepsilon_1(t)\varepsilon_2(t)] + M[\varepsilon_2(t)X(t)] + M[\varepsilon_2(t)\varepsilon_1(t)] + M[\varepsilon_2(t)\varepsilon_2(t)].$$

9.2 Technology for Monitoring the State of the Heart ...

Taking into account that

$$M[X(t)\varepsilon_1(t)] = 0, \quad M[\varepsilon_1(t)X(t)] = 0,$$
$$M[\varepsilon_1(t)\varepsilon_2(t)] = 0, \quad M[\varepsilon_2(t)\varepsilon_1(t)] = 0,$$

we have

$$D_g = M[X(t)X(t)] + M[X(t)\varepsilon_2(t)] + M[\varepsilon_2(t)X(t)] + D_{\varepsilon_1\varepsilon_1}$$
$$+ D_{\varepsilon_2\varepsilon_2} = D_X + 2M[X(t)\varepsilon_2(t)] + D_{\varepsilon\varepsilon}.$$

In view of the above, to perform control of the heart by means of smartphones, in addition to the normalized correlation technology, it is also advisable to use algorithms and technologies for calculating the spectral characteristics of the noise $\varepsilon(t)$ of the heart sound $g(t)$ from the expressions

$$a_{n_\varepsilon} = \frac{2}{N} \sum_{i=1}^{N} \operatorname{sgn} \varepsilon'(i\Delta t) \sqrt{|\varepsilon'(i\Delta t)|} \cos n\omega(i\Delta t),$$

$$b_{n_\varepsilon} = \frac{2}{N} \sum_{i=1}^{N} \operatorname{sgn} \varepsilon'(i\Delta t) \sqrt{|\varepsilon'(i\Delta t)|} \sin n\omega(i\Delta t),$$

In this case, the obtained estimates will differ from zero only in the presence of a correlation between the useful signal and the noise. Because of this, they allow registering the beginning of changes in the heart that precede the onset of a cardiovascular disease [1, 2].

In the following paragraphs, we present the results of a few of many computational experiments conducted in MATLAB to demonstrate the effectiveness of this technology. In the process of the experiments, we also calculated the estimates a_n, b_n, $\overline{\lambda_a(i\Delta t)}$, $\overline{\lambda_b(i\Delta t)}$, a_n^*, b_n^*, a_{n_ε}, and b_{n_ε} by means of the corresponding expressions given in Chap. 3.

During the experiment, heart sounds were analyzed daily in two stages. At the first stage, heart sound was recorded at rest. At the second stage, the subject was exposed to physical stress, immediately after which the heart sound was recorded. After this, using the described technologies, the robust spectral characteristics of the heart sound $g(i\Delta t)$ and the spectral characteristics of the noise $\varepsilon(i\Delta t)$ were calculated (Figs. 9.2 and 9.3).

At the first stage, the arrays of samples $\varepsilon'(i\Delta t)$ and $\varepsilon^e(i\Delta t)$ were formed. Then from the corresponding formulas, traditional estimates of the spectral characteristics a_n and b_n of the heart sound $g(i\Delta t)$ were calculated. After this, the estimates $\overline{\lambda_a(i\Delta t)}$ and $\overline{\lambda_b(i\Delta t)}$ and the estimates of the errors λ_{a_n} and λ_{b_n} were calculated. Further, the corrected estimates a_n^* and b_n^* were calculated. Finally, the estimates of the spectral characteristics a_{n_ε} and b_{n_ε} of the noise are calculated. In the course of the experiments, depending on the age group, for the spectra $n = 1 - 4$, the estimates a_n and b_n

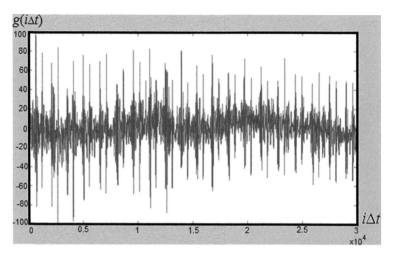

Fig. 9.2 Analyzed heart sound $g(i\Delta t)$

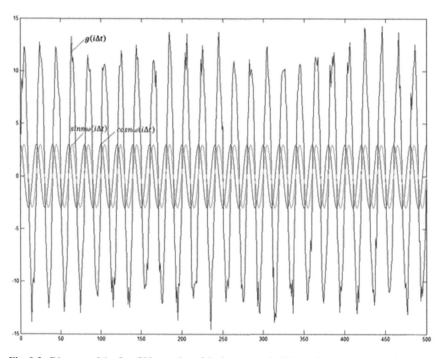

Fig. 9.3 Diagram of the first 500 samples of the heart sound $g(i\Delta t)$, sine waves and cosine waves

Table 9.7 Results of the computational experiment for Subject No. 1

State	D_ε	a_5	a_5^*	b_5	b_5^*	$a_{5\varepsilon}$	$b_{5\varepsilon}$
A.R.[a]	18.3	1.2	1.1	1.9	1.8	0.9	0.4
A.S.[b]	15.1	1.1	0.4	1.7	0.6	0.3	0.1

[a] At rest
[b] After stress

Table 9.8 Results of the computational experiment for Subject No. 2

State	D_ε	a_3	a_3^*	b_3	b_3^*	$a_{3\varepsilon}$	$b_{3\varepsilon}$
A.R.	17.1	1.8	1.6	1.9	1.9	1.1	0.7
A.S.	13.9	1.7	0.6	1.8	0.5	0.4	0.2

Table 9.9 Results of the computational experiment for Subject No. 3

State	D_ε	a_4	a_4^*	b_4	b_4^*	$a_{4\varepsilon}$	$b_{4\varepsilon}$
A.R.	15.1	1.5	1.5	1.6	1.7	1.2	0.6
A.S.	10.7	1.3	0.3	1.6	0.3	0.3	0.2

Table 9.10 Results of the computational experiment for Subject No. 4

State	D_ε	a_4	a_4^*	b_4	b_4^*	$a_{4\varepsilon}$	$b_{4\varepsilon}$
A.R.	16.2	1.8	1.7	1.9	1.7	0.9	0.4
A.S.	11.7	1.7	0.4	1.9	0.4	0.2	0.1

assumed values close to zero. Also depending on the age category, for $n = 3 - 5$, an abrupt jump was observed in the estimates a_n and b_n calculated from the traditional formulas.

As we can see from the results of the experiments, at $n = 3 - 5$, the estimates of the corrected spectral characteristics a_n^* and b_n^* of the noisy signal calculated from the proposed formulas were close to the values of the spectral characteristics calculated by traditional methods. The results of experiments reflecting this are presented in Tables 9.7, 9.8, 9.9, 9.10, 9.11.

As we can see from Tables 9.7, 9.8, 9.9, 9.10, 9.11, the estimates of the characteristics a_n^* and b_n^* of the heart sound at rest were close in value to the estimates of the spectral characteristics calculated by traditional technologies. However, after physical exertion, they reflect changes in the state of the heart more adequately than traditional estimates, which can also be observed through the changes in the estimates of the spectral characteristics of the noise a_{n_ε} and b_{n_ε}.

Table 9.11 Results of the computational experiment for Subject No. 12

State	D_ε	a_4	a_4^*	b_4	b_4^*	$a_{4\varepsilon}$	$b_{4\varepsilon}$
A.R.	10.2	1.1	1.2	1.5	1.5	0.8	0.6
A.S.	8.8	1.0	0.5	1.4	0.4	0.3	0.2

Therefore, we have reasons to believe that by using the algorithms for correcting the results of spectral analysis of heart sound and the technology for forming the informative attributes from spectral estimates of the noise in combination with traditional algorithms, we can improve the validity and reliability of the results of the monitoring of changes in the state of the functioning of the cardiovascular system [1, 2].

9.3 Correlation System for Monitoring the Beginning of the Latent Period of Vascular Pathology in Human Body

Vascular pathology is one of the most common diseases of human body. Vascular disorder occurs on different anatomical levels and affects various strata of society, evolving into a priority problem. Complexity of the problem is aggravated by the imperfection of methods of identification of damaged vessels at early stages. These methods basically allow one to solve the problem when a disease is already in its express form and development of vascular disorder reaches the level, at which disease is diagnosable by means of known methods but by that time its treatment becomes significantly complicated [1, 2, 27].

One of main sources of diagnostic information is rheovasosignals. They are periodic variables that characterize the condition of cardiovascular system. Noise of rheovasosignal is to a certain degree also a carrier of information on its condition. Presence of correlation between the useful signal and the noise of the rheovasosignal can also be used as an informative attribute about the beginning of the latent period of the ischemic disease.

Considered below is one of the possible forms of data interpretation of rheovasographic research, which allows one to a certain degree to solve the problem of identification of the beginning of the latent period of vascular disorder and deviation from the normal condition of cardiovascular system.

The distinctive feature of the method is application of a new diagnostic criterion in clinical practice allowing one to perform monitoring of the ischemic disease in the beginning of its onset.

Biophysical basics of the method are as follows. Abstracting the blood filling process, let us consider a model of a vessel with impaired patency, assuming that a blood vessel is a rigid tube, in which impaired patency is simulated by local hemodynamic resistance. Let us denote the cross-sectional area of a vessel in the local resistance area by S_0, and by S_1 for intact areas. Naturally, shapes and geometric dimensions of local resistance can be different in reality but they are mainly adequate in terms of hydrodynamics.

Reynolds number Re characterizing the blood flow in a vessel should be selected on the assumption that blood is a non-Newtonian fluid, for which values of effective (apparent) viscosity should be used. On the other hand, in modern literature on geodynamics, the authors of this book have not found values of numbers Re that correspond to the effectiveness of blood viscosity. Thus, an unambiguous choice of number Re is rather complicated for the proposed model. Considering the aforementioned and also taking into account the fact that blood flow in vessels is laminar, assume that Re < 25 in the model under consideration.

Our research demonstration that impaired arterial patency cause reduction of blood flow rate by the quantity

$$\frac{S_1}{S_0}\sqrt{1+\frac{25.2\sqrt{\frac{S_0}{S_1}}}{\text{Re}}} > 1.$$

Note that the effect of postarterial sections is not accounted for in derivation of this relationship. In reality, capillary vessels and veins have a significant effect on the nature of the blood flow in arteries. Therefore, this relationship defines only primary factors influencing blood flow in damaged vessels.

Accordingly, the value of rheovasosignal at the output of the sensor, which carries the information on the condition of all sections of the vascular bed, depends on the nature of blood filling process, i.e., the influence on the blood flow of factor S_0.

In other words, arterial inflow in damaged vessels is a function in luminal area S_0 of the vessel in the segment of local hemodynamic resistance.

Experimental study of numerous patients indicated that arterial inflow and venous outflow per time unit actually changes in such functional disorders, which causes not only to a change in signal amplitude but also a time shift between peaks of signals corresponding to damaged and undamaged vessels. Also, as the severity of malfunction increases, the time shift between the peaks grows. Hence, monitoring of the beginning of origin of vascular pathology of human body can be reduced to determining of the time shift between rheovasosignals received from symmetrical limbs of human. Shown in Fig. 9.4 is a technology for monitoring the functional state of limb vessels, using the delay time $\Delta \tau$ between the signals $g_v(t)$ and $g_j(t)$ taken by two pairs of electrodes from conjugate points of symmetric segments of a patient's limbs. The received rheovasosignals are used to calculate their cross-correlation function. Using the delay of the peak in time of $R_{g_v g_j}(\mu \Delta t)$, the severity of the patient's vascular disease is established. To obtained unambiguous estimates, the procedure is carried out in several reversible cycles and $\Delta \tau$—the lowest value of t—is chosen from the obtained results.

Figure 9.4 shows the flow chart of the system implementing the proposed method. Rheovasosignals are taken from corresponding points on symmetric areas of the patient's limbs by means of sensors of a cardiograph. After amplification, the obtained rheovasosignals are converted to digital codes. As a result, the samples of the rheovasosignals $g_v(i\Delta t)$ and $g_j(i\Delta t)$ are used to calculate the estimates of the cross-correlation functions $R_{g_v g_j}(\mu = 0\Delta t)$, $R_{g_v g_j}(\mu = 1\Delta t)$, $R_{g_v g_j}(\mu = 2\Delta t)$, ..., $R_{g_v g_j}(\mu = m\Delta t)$, whose curves are shown in Fig. 9.4.

If the patient is healthy, then the estimates of the cross-correlation function $R_{g_v g_j}(\mu)$ at $\mu = 0\Delta t$ assume the maximum value (curve 1 in Fig. 9.4). Otherwise, the larger $\Delta\tau = \mu\Delta t$, at which $R_{g_v g_j}(\mu)$ assumes the maximum value, the more alarming the patient's condition can be considered (curve 2 in Fig. 9.4).

In the general case, the differences of the time shifts $\Delta\tau_{12}$, $\Delta\tau_{13}$, $\Delta\tau_{14}$, $\Delta\tau_{23}$, $\Delta\tau_{24}$, $\Delta\tau_{34}$ between the rheovasosignals taken from four limbs of a patient, in the normal state of patient's cardiovascular system should be within the range established by highly experienced cardiologists. Otherwise, when these time shifts exceed the established range, patient is referred to the risk group. Since diseases of limb vessels, particularly of lower limb vessels, is one of the most complicated problems of general surgery and vascular surgery, we can assume that, this method, which does not require using expensive equipment, can be of practical use for mass medical examination of population [1, 2, 27].

In general case, the time shift $\Delta\tau_{ij} = \mu_{max}\Delta t$ between two correlated noisy random signals $g_v(i\Delta t)$ and $g_j(i\Delta t)$ can be calculated from the expression

$$R_{g_v g_j}(\mu\text{max}) = \frac{1}{N}\sum_{i=1}^{N} g_v(i\Delta t)g_j(i+\mu)\Delta t.$$

In that case, the time $\Delta\tau_{ij} = \mu_{max}\Delta t$, at which estimate $R_{g_j g_i}(\mu)$ assumes the maximum value, is determined on the basis of the results of the calculation.

Described are some alternative technologies for calculating the desired time shift $\Delta\tau_{vj}$:

$$R^*_{g_v g_j}(\mu) = \frac{1}{N}\sum_{i=1}^{N} g_v^2(i\Delta t)g_j^2(i+\mu)\Delta t,$$

$$R^*_{g_v g_j}(\mu) = \frac{1}{N}\sum_{i=1}^{N} g_v(i\Delta t)g_j^2(i+\mu)\Delta t.$$

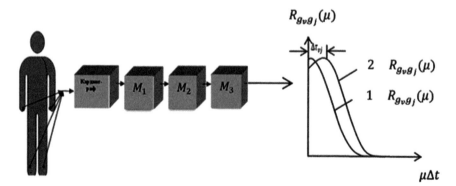

Fig. 9.4 Correlation system for monitoring the beginning of ischemic diseases: M_1—module of formation of samples of $g_v(i\Delta t)$, $g_j(i\Delta t)$; M_2—software module of correlation analysis; M_3—software module of calculating of $\Delta\tau$

9.3 Correlation System for Monitoring the Beginning of the Latent Period ...

The time shift $\Delta \tau_{vj}$ between the corresponding limbs, as shown in Fig. 9.4 is displayed on the computer monitor, which allows evaluating patient's condition visually.

Experimental research has demonstrated that the presence of the pronounced time shift $\Delta \tau_{ij}$ between the rheovasosignals $g_v(i\Delta t)$ and $g_j(i\Delta t)$ received from patient's different limbs is a reliable informative attribute of the monitoring of the beginning of a cardiovascular disease [1, 2, 27].

In conclusion, it should be noted that, as mentioned earlier, the beginning of the latent period of ischemic diseases can also be controlled using the estimate of the cross-correlation function between the useful signal $X(i\Delta t)$ and the noise $\varepsilon(i\Delta t)$ of the rheovasosignal. If patient's limbs are health, the noise $\varepsilon(i\Delta t)$ is formed only by the impact of external factors and therefore has no correlation with the useful signal $X(i\Delta t)$, i.e., $R_{X\varepsilon}(\mu \Delta t) = 0$. However, in vessels with impaired permeability, the noise $\varepsilon(i\Delta t)$ is correlated with the useful signal $X(i\Delta t)$ and the cross-correlation function, in this case, differs from zero [1, 2, 27]. Because of this, we can use the estimates of the relay cross-correlation function $R^*_{X\varepsilon}(\mu)$ to control the beginning of the latent period of ischemic diseases, calculating them from the expression

$$R^*_{X\varepsilon}(m) = \frac{1}{N} \sum_{i=1}^{N} \operatorname{sgn} g(i\Delta t)[g(i+m)\Delta t + g(i+m+2)\Delta t - 2g(i+m+1)\Delta t].$$

And the values of the estimates $R^*_{X\varepsilon}(\Delta t)$, $R^*_{X\varepsilon}(2\Delta t)$, $R^*_{X\varepsilon}(3\Delta t)$, ... allows us to control the degree and the stage of deviation of patient's limb vessels from the normal state. When $R^*_{X\varepsilon}(\Delta t) = 0$, there is no deviation. When $R^*_{X\varepsilon}(\Delta t) > 0$ and $R^*_{X\varepsilon}(2\Delta t) = 0$, a deviation starts, and when $R^*_{X\varepsilon}(\Delta t) > 0$ and $R^*_{X\varepsilon}(2\Delta t) > 0$, an obvious deviation takes place. In this case, patient must be diagnosed by traditional methods.

References

1. Aliev TA, Rzayeva NE, Sattarova UE (2017) Robust correlation technology for online monitoring of changes in the state of the heart by means of laptops and smartphones. Biomed Signal Process Control 31:44–51. https://doi.org/10.1016/j.bspc.2016.06.015
2. Aliev TA, Alizada TA, Rzayeva NE (2012) Noise control of heart by means of a smartphone. Lambert Academic Publishing, Saarbrücken
3. Holter monitor. https://en.wikipedia.org/wiki/Holter_monitor. Accessed 21 May 2018
4. EASY ECG Mobile. www.atesdevice.it/site/ecg/ecgmobile.html. Accessed 21 May 2018
5. Böhm C, Khuri S, Lhotská L, et al (2011) Information technology in bio- and medical informatics. In: Second international conference, ITBAM 2011, Toulouse, France, 31 Aug–1 Sept 2011. https://doi.org/10.1007/978-3-642-23208-4
6. Brause RW, Hanisch E (2000) Medical data analysis. In: First International Symposium, ISMDA 2000 Frankfurt, Germany, 29–30 Sept 2000. https://doi.org/10.1007/3-540-39949-6
7. Kącki E, Rudnicki M, Stempczyńska J (2009) Computers in medical activity. Springer, Berlin, Heidelberg. https://doi.org/10.1007/978-3-642-04462-5
8. Perry IR (1984) Real-time clinical computing. Research Studies Press, Letchworth

9. Rakus-Andersson E (2007) Fuzzy and rough techniques in medical diagnosis and medication. Springer, Berlin, Heidelberg. https://doi.org/10.1007/978-3-540-49708-0
10. Aliev TA, Zabirova AK, Mamedov ID (1982) A device for disease diagnostics. Author's certificate № 999064
11. Varghees VN, Ramachandran KI (2014) A novel heart sound activity detection framework for automated heart sound analysis. Biomed Signal Process Control 13:174–188. https://doi.org/10.1016/j.bspc.2014.05.002
12. Hoover A, Singh A, Fishel-Brown S et al (2012) Real-time detection of workload changes using heart rate variability. Biomed Signal Process Control 7(4):333–341. https://doi.org/10.1016/j.bspc.2011.07.004
13. García CA, Otero A, Vila X et al (2013) A new algorithm for wavelet-based heart rate variability analysis. Biomed Signal Process Control 8(6):542–550. https://doi.org/10.1016/j.bspc.2013.05.006
14. Berg M (1997) Rationalizing medical work: decision-support techniques and medical practices. MIT Press, London. https://doi.org/10.1086/383903
15. Bos L, Roa L, Yogesan K, O'Connell B et al (2006) Medical and care compunetics 3. IOS Press, Amsterdam
16. Cios KJ (2000) Medical data mining and knowledge discovery. IEEE Eng Med Biol Mag 19(4):15–16. https://doi.org/10.1109/MEMB.2000.853477
17. Ellis D (1987) Medical computing and applications. Ellis Horwood, Harlow
18. Helopoulos C (2003) The medical professional's guide to handheld computing. Jones and Bartlett Learning, Sudbury
19. Lazakidou AA, Siassiakos KM (2008) Handbook of research on distributed medical informatics and e-health. Medical Information Science Reference, New York
20. Lee N, Millman A (1995) ABC of medical computing: hospital based computer systems. BMJ, London. https://doi.org/10.1136/bmj.311.7011.1013
21. November JA (2012) Biomedical computing: digitizing life in the united states. The Johns Hopkins University Press, Baltimore
22. Schattner P (2007) Computing and information management in general practice. McGraw-Hill Education, New York
23. Schmitt M, Teodorescu H, Jain A et al (2002) Computational intelligence processing in medical diagnosis. Physica, Heidelberg. https://doi.org/10.1007/978-3-7908-1788-1
24. Tan J (2008) Medical informatics: concepts, methodologies, tools, and applications. IGI Global, Hershey. https://doi.org/10.4018/978-1-60566-050-9
25. Collacott RA (1989) Structural integrity monitoring. Chapman and Hall, London
26. Sakamoto Y, Ishiguro M, Kitagawa G (1986) Akaike information criterion statistics. Kluwer Academic Publishers, New York
27. Aliev T (2007) Digital noise monitoring of defect origin. Springer, Boston. https://doi.org/10.1007/978-0-387-71754-8
28. Aliev T (2003) Robust technology with analysis of interference in signal processing. Kluwer Academic/Plenum Publishers, New York. https://doi.org/10.1007/978-1-4615-0093-3
29. Aliev TA, Guluyev GA, Pashayev FH et al (2012) Noise monitoring technology for objects in transition to the emergency state. Mech Syst Signal Process 27:755–762. https://doi.org/10.1016/j.ymssp.2011.09.005
30. Aliev TA, Abbasov AM, Guluyev QA et al (2013) System of robust noise monitoring of anomalous seismic processes. Soil Dyn Earthquake Eng 53:11–25. https://doi.org/10.1016/j.soildyn.2012.12.013

Index

A
Accident development dynamics
   correlation technology of noise control, 33
Accident dynamics, 86, 88, 92, 94
   accident initiation and development
      sensors and specifics of information
      support, 6
Accident initiation, 1, 9, 12
Accidents, 1–3, 5–9, 12, 16, 20, 21, 48, 54
   at oil fields, 81
Algorithms
   beginning of the latent period of accidents, 64
Algorithms and technologies for calculating the errors, 59–62
   object's emergency state, 59-62
Analog-to-digital conversion, 75, 76
Analysis of heart sounds, 182, 183
Analysis of noisy signals, 36
Aviation equipment, 135, 136, 138

B
Bulla–Deniz offshore, 80

C
Cardiovascular system, 181, 194, 196
Center of monitoring of anomalous seismic processes, 154
Compressor stations, 101, 115, 118
Construction facilities, 155, 157
Controlling accident initiation
   correlation analysis, 13
Control of accident initiation
   spectral analysis methods, 16
Control of the beginning of defect initiation, 19

Control of the beginning of the latent period of accidents, 6, 8, 121, 133
   Position-Binary Technology (PBT), 68-72
Correlation, 27
Correlation analysis, 8, 13, 16, 27, 82, 92, 103, 105, 141, 183, 195
Correlation analysis of noisy signals
   latent period of transition, 27
Correlation matrices, 45, 47, 48, 50, 51, 53, 55, 56, 83, 88, 125, 126
   latent period of object's emergency state, 45–48
Correlation noise technology, 56
Correlation system, 194, 196
Correlation technology, 27
Cross-correlation functions, 37

D
Defect development dynamics, 33
Defects, 1–6, 11, 20, 33, 53, 73, 82, 85, 86, 115, 117, 122–124, 131, 133, 135, 136, 142
Density of distribution of noise, 40
   by its equivalent samples, 40
Developing noise controllers
   vibration sensors, 140
Development dynamics, 64–67, 77
Development dynamics of accidents, 37
Development dynamics of the latent period of accidents, 37
Digital citywide system, 155
Drilling accidents, 88, 91
Drilling rigs, 79–81, 89–93, 98, 99
Dynamics, 1, 2, 5, 9, 21
Dynamometer card, 102–108, 110, 113, 114

E

Effects of the signal filtration, 19
Effects of traditional methods
   sampling interval selection
      control of the beginning of accident initiation, 20
Emergency state, 122, 124, 126, 129, 131, 133–137, 139
Environmental pollution of soil, 88
Equivalent correlation matrix, 49, 50, 52
   latent period of object's emergency state, 52
Ericsson Mobile Health(EMH) system, 182
Estimate of noise variance, 27, 28, 33, 37

F

Filtration, 10, 20
Fixed offshore platform, 79, 88
Fixed platform (FP), 80
Formation of correlation matrices, 45, 46
FUGRO Offshore Structural Monitoring Systems, 81

H

Hazard warning network, 140
Hydroelectric power plants, 121

I

Identification of technical conditions, 101, 105, 107, 108, 114
Informative attributes, 101, 107–111, 113, 114, 118
Intelligent noise system, 121, 129

L

Laptops and smartphones in cardiology, 181, 182
Latent period of accidents, 6, 8, 10–12, 27, 29, 33, 37, 62, 64, 66–68, 73, 74
Latent period of initiation and development of accidents, 69
Latent period of object's emergency state, 59–65
Latent period of the accident, 60

M

Medicine, 181

N

Neft Dashlari, 79
Noise analysis, 8
Noise analysis of latent period, 45, 54
Noise analysis of seismic processes, 145–147, 150, 155, 157, 158
Noise analysis of vibration signals, 116
Noise control, 10, 27, 115, 116, 118, 184
Noise control of Ischemia, 194, 197
Noise control of vascular pathology, 195
Noise control system, 79, 82, 83, 88, 91–93, 97–99, 101
Noise control technology
   railway safety systems, 139
Noise diagnostics system, 131
Noise monitoring, 103
Noise monitoring of seismic stability, 145, 147, 156, 157
Noise monitoring of the state of the heart, 181, 185, 186, 188
Noise sampling, 59, 61–64, 70
Noise signaling, 27, 36, 68
Noise technology, 88
Noise variance, 90
Noisy signal, 29, 32, 34, 39, 41, 45, 47–49, 51, 54, 55
Nuclear power plants, 121–124, 129

O

Object state, 46
Offshore oil platform, 79
Oil and gas pipelines, 115
Oil field facilities, 108, 115
Online noise diagnostics, 129

P

Position-binary technology (PBT), 68, 69, 71, 73, 74
Possibilities of using laptops and smartphones control of the state of the heart, 181
Power engineering facilities, 121, 122, 124–127, 129
Preflight noise control, 137

R

Railway safety, 121, 139, 140
Rheovasographic research, 194
Rig control system (RCS), 92
RNM ASP station, 147–150, 154, 159, 160, 162–166, 175–178
Robustness, 104

S

Sampling, 13, 17, 21
Sampling interval of analog signals, 74
Seismic-acoustic signals, 145–148, 150, 152, 154, 161
Seismic-acoustic system, 145, 158, 176
Seismic stability, 145

Sensors, 1, 2, 6–8, 10, 12, 13
Signal model, 16
Signal processing, 86
    spectral analysis methods, 10
Spectral analysis methods, 10
Spectral characteristics in the latent period, 59
Spectral characteristics of heart sounds, 190
Spectral noise analysis, 59, 62, 64, 66
Spectral noise characteristics, 59, 62, 64, 66
Subsystem of noise control of malfunctions, 140
Sucker rod pumping units, 101

T
Technical facilities, 1, 3, 12
Technology and system of noise control power engineering facilities, 121
Technology for monitoring the state of the heart, 190
Traditional technologies, 9, 12, 13, 27, 54, 59, 88, 121, 122, 146, 153, 157, 159, 193

U
Useful signal, 45–47, 49–51, 53, 55, 56

V
Variance of noise of noisy signals, 27
Vascular pathology, 194
Vibration displacement sensors, 98
Vibration velocity analysis, 118

W
Wiener–Khinchin theorem, 19